SCIENCE LITERACY
for the
TWENTY-FIRST CENTURY

epilogue by Nobel Laureate **LEON LEDERMAN**

SCIENCE LITERACY *for the* TWENTY-FIRST CENTURY

contributors include:

Stephen Jay Gould

Howard Gardner

Margaret J. Geller

James Trefil

Sheila Tobias

Lawrence M. Krauss

George A. Keyworth

Bruce Alberts

edited by

STEPHANIE PACE MARSHALL,
JUDITH A. SCHEPPLER, & MICHAEL J. PALMISANO

Prometheus Books
59 John Glenn Drive
Amherst, New York 14228-2197

KH

Published 2003 by Prometheus Books

Inquiries should be addressed to
Prometheus Books
59 John Glenn Drive
Amherst, New York 14228–2197
VOICE: 716–691–0133, ext. 207
FAX: 716–564–2711
WWW.PROMETHEUSBOOKS.COM

07 06 05 04 03   5 4 3 2 1

Library of Congress Cataloging-in-Publication Data

Science literacy for the twenty-first century / contributors include Stephen Jay Gould . . . [et al.].; edited by Stephanie Pace Marshall, Judith A. Scheppler & Michael J. Palmisano ; epilogue by Leon Lederman.
    p. cm.
Includes bibliographical references.
ISBN 1–59102–020–4 (acid-free paper)
    1. Science—Study and teaching—United States. I. Gould, Stephen Jay. II. Marshall, Stephanie. III. Scheppler, Judith A. IV. Palmisano, Michael J., 1947-

Q183.3.A1 S3565 2002
507'.1'0073—dc21

                                                            2002036718

Printed in Canada on acid-free paper

10/25/04

Leon Lederman

CONTENTS

3. REFRAMING SCIENCE TEACHING

4. SCIENTIFIC STEWARDSHIP

5. BEYOND SCHOOLS: DEMYSTIFYING SCIENCE FOR PUBLIC POLICY

6. THE LEDERMAN LEGACY FOR EDUCATION

PREFACE

Science Literacy for the Twenty-first Century is a unique collection of essays written by noted scholars, educators, and scientists in honor and celebration of Dr. Leon Max Lederman. Lederman has long been a passionate advocate for science education, challenging educators, students, scientists, and policymakers for revolutionary change on its most fundamental levels. Through these provocative essays, all of us are challenged to think in transformative ways about our current educational system for science and the learning and teaching conditions to develop a more scientifically literate society.

ORGANIZATION

The contributing authors were invited to address issues of science literacy such as, What and how much science should everyone know? How should science be taught and experienced? How do we prepare science teachers? And, science as a human endeavor. The authors have written on a topic that they are passionate about. The essays naturally clustered into the critical issues of access and opportunity, reframing learning and teaching, stewardship, and public policy. We invite you to explore the issues through the voices of the scientists, scholars, and educators who contributed to this book.

Part 1, "Extending Invitations to Scientific Study," addresses the need for ensuring that a solidly grounded science education is available and accessible to all individuals. Not only is the knowledge of science needed but also more students, especially those traditionally underrepresented in the sciences, must be invited to the study of science. Parts 2 and 3, "Reframing Science Learning" and "Reframing Science Teaching," offer ideas for making the science that is taught more meaningful to the learners. While *what* is taught is critical, so, too, *how* science is taught determines our students' future interest and success in science. Part 4 addresses "Scientific Stewardship." Part 5, "Beyond Schools: Demystifying Science for Public Policy," addresses the issue of the public understanding of science, underscoring that responsibility for literacy in science does not reside solely with our educational system but also with scientists and policymakers.

Part 6, "The Lederman Legacy for Education," provides a glimpse of three very different learning enterprises that exist today, largely because of Leon Lederman's vision, tenacity, and influence. He has placed his indelible stamp on these institutions for the advancement of scientific literacy. These learning entities are exemplars of the critical dimensions of improved scientific literacy: access and invitation, transformational teaching and learning, stewardship, and attention to public policy.

Science Literacy for the Twenty-first Century was conceived as a surprise birthday gift to Leon Lederman on his eightieth birthday. Busy schedules and the business of publishing conspired so that while the manuscript for the book was finished well before Leon's birthday celebration, the book was actually not due in published form for several months after the celebration. This, and the flexibility of Prometheus Books, allowed us the opportunity to add one final voice, that of Leon Lederman himself, who speaks so eloquently and passionately to the need for a scientifically literate public. While this may seem a bit curious, asking one to contribute to one's own gift, the gift of this book to its readers and our society seemed incomplete without his humor and wisdom. Throughout the book, you will also find whimsical drawings created by Leon Lederman for his many talks in support of science literacy.

ACKNOWLEDGMENTS

We would like to thank our contributing authors for their gifts of time and expertise. They are truly some of the top science leaders of our time and yet they generously offer us their wisdoms and insights.

No book of this magnitude is possible without the assistance of many individuals. We would first like to thank Denise Koehnke and Kathy Thulis for their expert assistance. At Prometheus Books, we would like to thank our acquisitions editor Linda Greenspan Regan, production editor Christine Kramer, and art director Jacqueline Cooke for their patient assistance.

Stephanie Pace Marshall
Judith A. Scheppler
Michael J. Palmisano

INTRODUCTION

Stephanie Pace Marshall, Judith A. Scheppler, and Michael J. Palmisano

The twenty-first century finds science progressing at a rate that increasingly outpaces the wisdom necessary for using newfound knowledge for public good and sustainability of our global world. This raises critical questions concerning essential understandings of science for all citizens. National security, economic viability, and the health and welfare of families and communities all require increasingly deeper levels of understanding of science. Can public education foster science literacy for all?

While education reforms of the past decade have produced incremental results, their promise of ensuring that all students acquire essential knowledge and skills is far from realized. The performance of U.S. students on international assessments of science knowledge and skills is disappointing and unacceptable. Too many students are unable to meet the expectations specified in state and national science learning standards and greatly disproportionate numbers of minority and poor students do not meet the standards.

Adding to the problem, current indicators of achievement and success—state assessments, norm-referenced exams, and college admissions tests—are inadequate measures of science literacy. While they emphasize acquisition of content knowledge and skills, little regard is given to knowledge generation, evaluation of information, and ethical applications of knowledge to real-world problems and issues. Also absent is assessment of inquiry and exploration that are essential to science. Even for those students who perform well on current measures of achievement and go on to earn college degrees, mounting evidence suggests that they graduate with thinking rooted in naïve misconceptions of the natural world (Schneps and Sadler 1987; Gardner 1991; Perkins 1992).

What and how much science everyone should know is of great debate. Science curricula are bloated with facts and information and often neglect the need to foster the ability to evaluate and apply information. Cultivation of the curiosity and skeptical inquiry necessary for a scientific cast of mind is sacrificed in favor of broad and superficial understanding. Textbooks that often lack currency, depth, focus, rigor, and accuracy drive curricula. The Third International Mathematics and Science Study (TIMSS) characterized textbooks as "a mile wide and an inch deep" (National Research Council 1996). The effect of state and national reforms has been to add even more content to traditional topics, thereby compounding the problem and diminishing the opportunity to engage in science exploration.

The student achievement problem is exacerbated by the issues of teacher shortage and quality. The long-standing need to increase the numbers of well-qualified teachers of science remains unfulfilled. More than one-third of America's students are taught science by teachers who do not meet minimum certification requirements. The situation is worse in urban and rural schools where more often poor and minority students are taught mathematics and science by less qualified teachers. Moreover, there simply are not enough students of science in the pipeline to meet the demand for qualified science teachers.

The call to action for increased science literacy is well stated in *A Report to the Nation from the National Commission on Mathematics and Science Teaching for the Twenty-first Century*:

> In short, our children are losing the ability to respond not just to the challenges already presented by the twenty-first century but

to its potential as well. We are failing to capture the interest of our youth for scientific and mathematical ideas. We are not instructing them to the level of competence that they will need to live their lives and work at their jobs productively. Perhaps worst of all, we are not challenging their imaginations deeply enough.[1]

It is within this context that we pose these two questions: *If anything imaginable were possible, if there were no constraints whatever, (1) what would it take for all students to acquire a strong base of content knowledge and skills, the skills to acquire and generate new knowledge, and the skills to evaluate and apply knowledge to academic and real-world issues? (2) What would it take to rekindle students' innate curiosity and to nourish their predisposition for exploration and discovery?*

In order for students to acquire these capacities, policymakers, educators, scientists, and the general public will need to commit to fostering learning outcomes that cultivate deep understanding and the skills and predispositions necessary for the scientific cast of mind. The prevailing emphasis on passive acquisition of knowledge and competition among learners must be reframed. Learning must be recognized as active, volitional, enhanced by challenge, and inhibited by threat. Priority must be given to deep conceptual understanding of important ideas, the ability to represent knowledge, and to integrative thinking. Students need firsthand experience with inquiry, reasoning, experimentation, data collection, analysis, and truth verification. They need to experience interdisciplinary learning and applications of knowledge to real-world situations so they come to understand themselves, their world, and their place in it. In short, schools will need to fundamentally change the way students and teachers engage in science and with each other so that all students build a foundation of science understanding to serve them as learners, leaders, workers, and citizens—regardless of their life choices in the future.

Schooling will change when the public, policymakers, and educators understand and support the need for cultivating the scientific cast of mind for all students. In Leon Lederman's own words:

We are living through a time of profound change in our world. After September 11, 2001, ordinary activities have taken on a new meaning and a new significance. One of these goes under the broad category of education. Our world is driven by scientific and technological change. We are all too familiar with science's fundamental contributions to our society—economic growth, transportation, communications, nutrition, health, entertainment and the easy availability of information. Policymakers and citizens appreciate these contributions, but do they understand that science proceeds by trial and error, by the involvement of talented young people who have been advantaged by good education and by an infrastructure that has taken generations to construct? Are they—indeed is the educated public, including the scientists, media, and policymakers—aware of the complex effect of new knowledge and innovation, whose consequences are often unexpected and even disorienting?[2]

Democracies and the global community require scientifically literate citizenries. The scientists, scholars, and educators writing for this book have been challenged to provoke our thinking. They address the need for increased access and opportunity for scientific study for all students. They grapple with the need to reframe the science curriculum and the teaching of science. They speak to the need for scientific stewardship and public policy implications for science literacy. And they describe learning institutions fostered and inspired by Leon Lederman and his vision for science literacy for all.

Together, they embrace science as a human endeavor and understanding science as a human imperative.

NOTES

1. U.S. Department of Education, National Commission on Mathematics and Science Teaching for the Twenty-First Century, *Before It's Too Late* (Washington, D.C.: U.S. Government Printing Office, 2000).

2. Statement to the press by Leon Lederman on November 14,

2001, for a book launch of *Portraits of American Scientists* authored by students of the Illinois Mathematics and Science Academy and edited by Leon M. Lederman and Judith A. Scheppler.

REFERENCES

Bass, H., Jane B. Kahle, et al. *Mathematics and Science Education around the World: What Can We Learn from the Survey of Mathematics and Science Opportunities (SMSO) and the Third International Mathematics and Science Study (TIMSS)?* Washington, D.C.: National Academy Press, 1996.

Gardner, Howard. *The Unschooled Mind: How Children Think and How Schools Should Teach.* New York: Basic Books, Inc., 1991.

Perkins, D. M. *Smart Schools: From Training Memories to Educating Minds.* New York: The Free Press, 1992.

Schneps, M. H., and P. M. Sadler. *A Private Universe.* South Burlington, Vt.: Annenberg/Corporation for Public Broadcasting Multimedia, 1987.

THE DESIGN OF THE NEW HIGH SCHOOL
PROGRAM IS FOR ALL STUDENTS

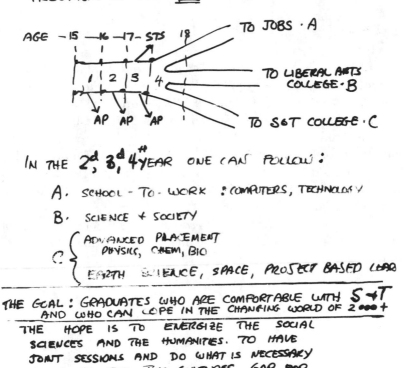

AGE — 15 — 16 — 17 — STS 18

TO JOBS · A

TO LIBERAL ARTS COLLEGE · B

TO S&T COLLEGE · C

AP   AP   AP

IN THE $2^{d}, 3^{d}, 4^{\#}$ YEAR ONE CAN FOLLOW:

A. SCHOOL-TO-WORK : COMPUTERS, TECHNOLOGY

B. SCIENCE & SOCIETY

C. { ADVANCED PLACEMENT
PHYSICS, CHEM, BIO

EARTH SCIENCE, SPACE, PROJECT BASED LEARN

THE GOAL: GRADUATES WHO ARE COMFORTABLE WITH S&T
AND WHO CAN COPE IN THE CHANGING WORLD OF 2000+

THE HOPE IS TO ENERGIZE THE SOCIAL
SCIENCES AND THE HUMANITIES. TO HAVE
JOINT SESSIONS AND DO WHAT IS NECESSARY
TO CLOSE THE TWO CULTURES GAP FOR
THE 'ARISE' GENERATION

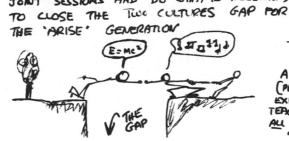

$E = mc^2$

THE
GAP

THE HIDDEN
AGENDA

A HIGH SCHOOL
[PERHAPS 8-N]
EXPERIENCE THAT
TEACHES THE UNITY OF
ALL KNOWLEDGE:
" CONSILIENCE.

PART 1
EXTENDING INVITATIONS TO SCIENTIFIC STUDY

Opening the Doors of Science

Margaret J. Geller

In 1995, the Ford Foundation sponsored a series of public lectures on science. Along with the lecture at the 92nd Street Y in New York City, the organizers asked each speaker to have dinner with a large group of high school students from special city-sponsored math and science programs and to do a radio interview on the oldest public radio station in the United States, WNYC. My first order of business on the morning of the lecture was to get a genuine New York bagel; the taste and texture are just not the same anywhere else. Bagel in hand, I went to the studios of WNYC near City Hall. The street around City Hall was filled with demonstrators. That day, the city government had announced cuts in the education budget. Students, teachers, and other concerned citizens were wearing black ribbons to mourn the loss.

Leonard Lopate interviewed me on his WNYC program, *New York and Company*. With his consummate skill as host, he guided our conversation smoothly around the universe from an outline of the hot big bang model to the story of my own work mapping out the distribution of galaxies over distances of hundreds of millions of light years. During a station break, Lopate asked if there was anything he had omitted. I suggested that he ask me about education.

On the air, Lopate asked why I cared so passionately about public education in New York City when my life is "so removed" from it and when I come from a "family of scientists." I replied that I have a strong connection because my father went to Stuyvesant High School in New York City. His education at Stuyvesant was the foundation for his later success as a scientist. I continued that I was more concerned about budget cuts for public education than I was about limits on the budget for scientific research. Science is forward looking. Without solid, inspired investment now, which opens the doors to creative careers (including careers in science) for all young people, scientific research, as we know it, won't continue long into the future. Lopate responded that my scientific colleagues might be unhappy with my putting education above research as a priority for public funding. I replied that the United States is certainly rich enough to afford both.

Leon Lederman has devoted many years to making sure that the United States affords both excellent education and leadership in scientific research. His career is an example of affording both: Lederman's contributions to public education are as great as his contributions to science. As Lederman has recognized so well, for young people who want to become scientists, particularly physical scientists, the die is nearly cast by the time the student graduates from high school. Students from the best high schools often have more solid training in mathematics, the language of science. Students from high schools weak in mathematics and science are often aware of the weakness. In some of the most competitive colleges and universities, perhaps unintentional slights enhance these feelings of inadequacy. Comments from well-prepared students: "I did this stuff in high school. Didn't you?" Or from instructors, "You should have seen this stuff in high school." These signals about "inferior" preparation discourage some students who would otherwise go on in science and are a significant hurdle for others. Several outstandingly able students who took my introductory astronomy/physics course told me that comments from other students and faculty in mathematics and the physical sciences made them feel handicapped merely because they did not come from an outstanding high school. For these students, I can attest that the feeling of inadequacy did not correspond at all to their abilities, but it was impossible to convince them that a door, which appeared closed, was actually open. Why pursue these fields, they wondered, when there are areas where they feel welcome?

The quality of high schools is tightly tied to the economic level of the school district. Studies show that education in mathematics and in the physical sciences is particularly sensitive to the wealth of the school.[1] The largest differences in teacher preparation between rich and poor schools are in mathematics, chemistry, and physics. The enormous disparities among high schools amplify rather than mediate the effects of differences in parental education and economic level. Because economic and ethnic/cultural lines are often correlated in the United States, inequities in the quality of high schools are a contributor to the differences between the population of scientists and the population of the country. Physical scientists are mostly white and mostly male.

To provide better opportunities to some of the most talented students independent of the quality of their neighborhood high school, some major cities have long supported special science-oriented high schools with entry by examination. New York City has three of these schools (including Stuyvesant), which boast a daunting list of distinguished scientists as alumni. The Illinois Mathematics and Science Academy (IMSA), to which Leon Lederman has devoted many inspired years, builds on and broadens the reach of these specialized city schools. IMSA is an important step because it draws students from the entire state. It provides access for students from diverse areas, which, for one reason or another, cannot or simply do not support excellent science education. Every state should have at least one high school equivalent to IMSA not only for science but also for the arts. Existence of these schools might inspire other distinguished people to devote their later years to education in rich and unusual ways.

Critics of "exam" schools have questioned the wisdom of using exam performance or even performance on an exam coupled with other more qualitative criteria to determine admission to these special high schools. Others stand by rigid quantitative selection criteria. Some of the problems of using exams and even of requiring that the student take the initiative to apply are obvious. Students in inferior elementary or junior high schools are unlikely to perform as well on the exams as those from better schools. The relative performance may not reflect their abilities but just the difference in preparation. Sometimes parents can make up for deficiencies in the schools, but the ability of parents to do so is unfortunately correlated with education and economic level and thus probably with the quality of the local schools.

Likelihood of poor performance on exams might discourage advisors in poorer schools from suggesting that their best students apply to specialized academic high schools. Parents of children who attend these schools may be less likely to know about opportunities like IMSA. Highly educated, economically secure parents are likely to insist that the elementary schools prepare their children to compete successfully for admission to the best high schools and they encourage their own children to apply. They may even enroll their children in special courses to train for the entrance exams. It seems that probable biases in the segment of the applicant population might result in perpetuation of many of the traditional barriers to scientific careers. These barriers are more serious for students from segments of the population traditionally underrepresented in the scientific profession.

Knowing, wanting, and succeeding are three key words in attracting diverse students to specialized academic high schools. First, there is *knowing* that the schools exist. Today one of the prime paths to discovery is the Internet. Out of curiosity I checked the content and feel of the Web sites for two dozen public science high schools (http://www.mit.edu:8001/people/belville/ncsssmst.html). I also nosed around the sites for two private schools in the Boston area, which have excellent science programs, Philipps Andover Academy (http://www.andover.edu) and Milton Academy (http://www.milton.edu). Overall, I was surprised to see the marked contrast between the sites of the private and public schools in this list. The sites of many of the public schools are quite cold. For example, most come up with a picture of the physical plant rather than of a diverse student/faculty population. Among the public schools, I found the North Carolina School of Mathematics and Science (http://www.ncssm.edu) most enticing. It sends an inclusive, inviting message, which is not at all intimidating. It suggests that interested students contact the student ambassador, a wonderful and insightfully encouraging idea.

The two private school sites are similar in feel but fancier. They come up with pictures of people; the sites are rich in professional photos of warm interactions among faculty and students. People, especially young people, are mostly interested in other people. Even now I am drawn more to pictures of people enjoying what they are doing than to uninhabited facilities or to technical stunts.

One might argue that the underlying difference in these Web sites is the wealth of the school, but the site for the North Carolina School shows that a simple, well-conceived site can be as appealing as the ones where more money was probably spent. The real differences among the sites are in the subliminal messages they carry. A homepage that carries an immediate and strong visual message that diverse students are welcome might have the desired return. A site that is a bit informal (as young people are) and communicates that the faculty care about the students as individuals is at least worth the experiment.

After *knowing* comes *wanting* to be part of a vibrant academic community. For many students the whole idea is foreign to their experience at home or at school. Students have to be able to picture themselves at a place like IMSA or Bronx Science, or for that matter, Milton Academy. Something has to spark their interest, a teacher, a parent, a friend, an unusual contact or experience. Sometimes the inspiration comes from prolonged contact with people who have lived the experience. Sometimes a brief interaction awakens or simply locks into an inner need.

I have been amazed at the number and range of unpredictable "awakening" events people have reported to me. With all the planned activities surrounding education, it is often an outside the box event that brings an inner drive to the surface. I suppose that my father's story should have prepared me for the many quirk-of-fate stories I have heard. But, of course, I thought his story from the days of the Great Depression could not be a parable for today. I retell it here because I realize that his story is the story of some young people today. The details are different, but the theme is the same.

My father is a first-generation American who grew up on the Lower East Side of Manhattan and later in Sheepshead Bay. As a kid he wore out his left shoe riding his scooter to the Sheepshead Bay Branch of the New York Public Library. A local druggist and the branch library nurtured his fascination with chemistry. He was a bright, curious kid attending an inadequate junior high school. As my father describes it, he was in his mother's knitting shop one day when his cousin arrived. He had not seen his cousin for years. His cousin was going to Stuyvesant High School and told my father the school had advanced chemistry courses. Teachers at my father's junior high school may have known about Stuyvesant, but they didn't tell him. When he wrote to find out about admission, Stuyvesant told him he would have to take an exam. He never took it. Before the exam date, my father received a letter saying that his mathematics grades alone were good enough for entry.

My father went to Stuyvesant, but never took the advanced chemistry courses because there was too much else that fascinated him in the school's wide world. He did become a distinguished solid-state chemist and says, "I shudder to think what would have happened to me if my cousin had not told me about Stuyvesant."

In my father's story, there are some lessons about succeeding. Students can't succeed if no one informs them of the opportunities. How many bright students are there in inadequate schools whose parents, teachers, and guidance counselors can't or won't help them to formulate the dream of going to an outstanding school and to pursue the dream? The second lesson I take from the story is about opening the doors to success. Someone at Stuyvesant recognized, without the entrance exam, that my father was unusual and admitted him.

The story has made me wonder whether today doors could be

opened by simply offering a few of the best math and science students in disadvantaged schools admission to magnet high schools based simply on sustained academic performance. Interviews are a tool for confirming the talent and drive of the students. For these schools, an experiment geared to start the "tradition" of sending students to math and science high schools might inspire both teachers and students. There is abundant evidence that challenging goals and first successes often make substantial changes in people's perception of themselves and their opportunities. Some successful students from inadequate or disadvantaged elementary and junior high schools would probably enjoy serving as student ambassadors.

High school is a beginning and a foundation. Opening doors to attendance at the best high schools is an important step toward opening the doors to the scientific profession. The change in the composition of the scientific community is frustratingly slow. Despite the pressure for and discussion of women in science, there is shockingly little improvement in their presence on faculties in the first-tier institutions, in the academies, and on lists of prizewinners (particularly for the most distinguished prizes including the Nobel Prize). The situation for scientists of color is worse.

Young people need to see a diverse scientific community to be attracted to it. Science needs a diverse community because it thrives on the widest range of talents and perspectives. Its support depends on public trust, confidence, and possibly some identification with people who do science and some understanding of science as a human endeavor.

Most young people never have the opportunity to share their curiosity and dreams with a scientist. As an ambassador of science, Leon Lederman has carried the messages of the humanity of science and the thrill of discovery to many young people. His dedication will, I hope, encourage people who have been successful in a broad range of creative careers to follow his example. Their sustained presence and caring will undoubtedly produce many "awakening" experiences.

NOTE

1. The association of economics and school quality has been extensively studied. More information can be found in the following recent documents: *The State of America's Children* (1998) by Children's Defense Fund, available at http://childrensdefense.org; *Report on the Condition of Education* (1997) by the U.S. Department of Education; and *Years of Promise: A Comprehensive Learning Strategy for America's Children* (1996) by Carnegie Corporation of New York, available at http://www.carnegie.org.

WOMEN AND PHYSICS, PHYSICS AND WOMEN

A Puzzlement

Sheila Tobias

There can be no science literacy in the population absent the full participation of women in all fields of science, both because they represent 50 percent of the population, and because—as wives, mothers, teachers, writers, and public figures—they exercise a profound and increasing influence over everyone else.

In their June 2000 report *Women in Physics*, coauthors Rachel Ivie and Katie Stowe document that although an increasingly large number of girls have some exposure to physics in high school, and women are now earning more than half of all bachelor's degrees in the life sciences and half the baccalaureates in chemistry, U.S. physics continues to be the last frontier for women and girls. Moreover, as the authors conclude, no one appears to know why. "It is possible," they write, "that women still experience subtle discrimination leading them away from physics, and that women choose careers that are less clearly linked to physics."[1] There has been progress. By the end of the 1990s, twenty departments of physics—not at women's colleges—were graduating some female physics majors (between five and twelve per year; compare Table 2 with Table 1). Salary discrepancies at colleges and universities have ceased to

Table 1: Total for All Physics Bachelor's Degrees Awarded to All Genders

	Academic Institution	1994	1995	1996	1997
1	Massachusetts Institute of Technology	71	61	47	58
2	Harvard University	45	53	55	52
3	University of Washington–Seattle	31	42	34	42
4	California Institute of Technology	24	25	24	33
5	Rutgers the State University of New Jersey–New Brunswick	26	32	32	31
6	University of California–Berkeley	33	35	26	30
7	UCLA	32	31	36	29
8	Brigham Young University, Main Campus	33	37	42	29
9	University of Texas at Austin	29	30	36	27
10	University of North Carolina at Chapel Hill	27	29	30	27
11	United States Naval Academy	28	30	26	27
12	University of Chicago	29	22	20	24
13	University of California–San Diego	34	37	46	24
14	University of Minnesota–Twin Cities	24	26	25	23
15	Georgia Institute of Technology, Main Campus	40	37	39	23
16	University of Virginia, Main Campus	13	21	22	22
17	University of Michigan at Ann Arbor	22	24	23	22
18	University of California–Santa Cruz	25	23	22	21
19	University of California–Irvine	19	28	18	20
20	College of William and Mary	17	27	18	20
21	University of Florida	14	13	6	19
22	Texas A&M University, Main Campus	20	13	9	19
23	North Georgia College	14	9	9	19
24	University of Colorado at Boulder	26	25	20	18
25	Rensselaer Polytechnic Institute	29	23	36	18
26	Reed College	8	16	11	18
27	Pennsylvania State University–University Park	19	34	24	18
28	Longwood College	10	7	6	18
29	Illinois State University	15	19	18	18
30	University of Maryland–College Park	28	20	13	17
31	Purdue University, Main Campus	26	20	15	17
32	North Carolina State University–Raleigh	23	14	13	16
33	Harvey Mudd College	25	29	22	16
34	Carnegie Mellon University	24	20	20	16
35	Case Western Reserve University	9	12	16	16
36	Cornell University, All Campuses	18	25	22	16
37	Whitman College	3	13	5	15
38	Wesleyan University	8	15	13	15
39	SUNY College at Geneseo	12	14	15	15
40	SUNY at Albany	18	15	9	15
41	Virginia Polytechnic Institute and State University	23	25	13	14
42	Arizona State University, Main Campus	13	13	15	14
43	United States Air Force Academy	56	15	55	14
44	Rice University	9	24	14	14
45	Princeton University	28	11	18	14
46	Dartmouth College	16	11	11	14
47	University of Utah	10	15	13	13
48	University of Missouri–Rolla	8	5	6	13
49	University of California–Davis	23	16	11	13
50	Bates College	10	3	9	13

Table 2: Female Physics Bachelor's Degrees Awarded

	Academic Institution	1994	1995	1996	1997
1	Massachusetts Institute of Technology	13	12	11	12
2	Bryn Mawr College	10	6	9	11
3	Harvard University	9	13	11	10
4	University of Washington–Seattle	3	6	5	9
5	Rutgers the State University of New Jersey–New Brunswick	2	6	2	9
6	Mount Holyoke College	8	4	10	9
7	Whitman College	1	5	2	8
8	Smith College	3	6	4	7
9	University of North Carolina at Chapel Hill	8	6	7	6
10	University of Michigan at Ann Arbor	2	4	5	6
11	University of Chicago	5	4	2	6
12	Brigham Young University, Main Campus	5	7	13	6
13	Alabama A&M University	0	4	0	5
14	University of Colorado at Boulder	9	5	6	5
15	Texas A&M University, Main Campus	6	0	1	5
16	Southern University A&M College at Baton Rouge	1	1	5	5
17	Purdue University, Main Campus	4	5	2	5
18	Barnard College	1	3	2	5
19	College of William and Mary	3	6	4	5
20	Williams College	1	2	0	4
21	Wellesley College	4	4	6	4
22	University of Virginia, Main Campus	4	3	6	4
23	University of Minnesota–Twin Cities	4	2	6	4
24	University of California–San Diego	4	8	3	4
25	Susquehanna University	0	0	1	4

be statistically significant. Yet women remain "sorely underrepresented" in physics; they earn less than one-fifth of bachelor's degrees and only one-eighth of Ph.D.s in physics. This is in comparison with countries such as Argentina, Italy, and the Philippines, where women's representation in physics is far greater than here.

In the words of the fictional King of Siam, unable to comprehend Anna, the English governess he hired in to educate his children, "It's a puzzlement." When one rereads the books and articles written in the 1970s about the absence of women in science, the data and their interpretation appear dated. But if one substitutes "physics" for "science," everything is still all too true. Why is it that physics appears to be out of reach for highly able U.S. women, even those who show an early talent and affection for mathematics? Or, to say it differently, why is it that the U.S. physics community is so resistant to mainstreaming itself?[2] And why are physicists, if they are so able to probe the secrets of the universe, unable to comprehend how to address the gender skew?

In a wide-ranging, if not statistically grounded, set of inquiries, this author is querying by way of e-mail young (forty-five and under) women who bucked the trend: bachelor's, master's, and by far the majority of respondents, Ph.D.s in physics and astrophysics, who, as their responses indicate, were attracted to physics and continue to enjoy the field for the same reasons as their male age-mates. ("I came to college wishing to do pre-med," reports a woman Ph.D. in thin film physics, now working largely in physics education research. "I took my first physics course because it was required. But I loved it despite the fact—maybe because of the fact—that it was hard.") Yet, even when they are successful as ordinarily measured, these women do not thrive at the same rate and to the same extent as men. Their problems cluster around three issues: the culture of physics, which remains overmasculinized, some say a quasi-religion, and to their minds destructively competitive,[3] even misogynistic;[4] work-life issues that penalize people (not just women) who have working partners and children; and perceived abuse, which ranges from being discounted and not taken seriously by professors, employers, and colleagues to being openly discouraged, disliked, and having one's career intentionally derailed. These are strong sentiments on the part of the respondents. Only when I promise absolute anonymity do they emerge.

One explanation for the perceived abuse, indirectly offered by physicist Gerald Holton and sociologist Gerhard Sonnert in their 1990s study of 699 men and women scientists who received NSF postdoctoral awards is that women in physics remain a sub-critical mass.[5] This has three effects: even—especially—the ablest experience isolation, a feeling of being different, of being the "only girl in class." Judy Franz, now executive director of the American Physical Society, remembers being excluded from late-night physics study sessions when she was an undergraduate at Cornell University, not because of outright discrimination but because she (and not her male classmates) had to abide by parietal rules. A second, continuing effect on professionals in a field that is undersupplied with women is the disproportionate demands made on their time—a factor that emerges out of my interviews. A third effect is typical of the disadvantages that accrue to any minority group lacking in power; they don't have access to the resources they need to help themselves or to help others like themselves.

The kind of parietal rules that locked Judy Franz out of study groups are long gone, but what remains is a culture of exclusivity that even "those of us who made it in science," as Rita Colwell writes in her Introduction to Elga Wasserman's new study of National Academy women (*The Door in the Dream*), "feel the sting of prejudice."[6] Colwell is the current director of the National Science Foundation, former president of the University of Maryland (UMD) Biotechnology Institute, and is currently on leave from UMD. As a student, Colwell found her ideas "not taken as seriously as the ideas of the aspiring male scientists around me," was told during her graduate studies that professors "didn't want to waste a fellowship on a woman," and watched many of her ideas simply "hijacked" along the way. The second of the barriers she recalls would be illegal today. But what "law" can prevent senior professors from taking junior women's ideas seriously? or colleagues from "hijacking" a woman scientist's work?

Lest one think this is history, a woman astrophysicist holding a professorship at a prestigious university and still under forty-five years of age writes:

> In my career, I've encountered repeatedly, a profound absence of respect for my abilities and for my accomplishments. I've had people say to my face that I'm a difficult person, because I am reaching "above myself" in aspiring for membership on a prestigious committee, for example, and even worse. I've been ignored, as if I'm invisible in arenas where I know more about the subject on the table than any of the other participants. . . . I still don't feel at home in my own profession. And I'm one of the lucky ones. I have a good job. I've been "let in the door."

Why do otherwise rational scientists downgrade women's ability to do physics? It may be solipsism: extrapolating from their own narrow experience with women. Women physicists marry men in science.[7] But many men in physics do not marry women professionals. Their own wives' (and some, though not all, of their daughters') aversion to physics supports their views, and, from their intimates, they assume all women are alike. Of Sonnert's and Holton's high-achieving sample of scientists, 62 percent of the women doctorates married men with science doctorates, but only 19 percent of the men chose mates with those credentials. This may be changing, but the physicists in positions to hire, fire, reward, and punish younger women in their field are of an earlier generation and it is they who continue to set the tone, the rules, and the style.

Why the abuse? This is more complicated.

"If you'd asked me a year ago," reports one senior member of a large physics department in a land grant university, "I would have reported that we had three women professors in our department. This year, there are none. One left for personal reasons, one to follow a husband to another part of the country, the third because she couldn't take it any more."

The Massachusetts Institute of Technology (MIT) study of the status of women faculty documented discrimination not just in the hiring of younger faculty and research associates but something *more than discrimination* in the treatment of senior women, the distribution of awards and promotions, the allocation of lab and office space, and their underrepresentation on important committees.[8] The committee's key finding, writes Elga Wasserman in her interpretation of the MIT study, is this: "Each generation of young women began by believing gender discrimination had been 'solved,' yet with senior status, they found the playing field not equal."[9] Indeed, the senior women among the National Academy members whom Wasserman interviewed echo the sentiments of the younger women in physics profiled in my study. From astronomer Vera Rubin, "Science is still a male-dominated profession and some of the males enjoy this dominance."[10] From Myriam Sarachik, an experimental physicist, "the degree to which your being a woman colors their view of how good you are."[11]

Indeed, in a 1995 post-award study of applicants for postdoctoral fellowships awarded by the Swedish Medical Research

Council, women needed more than one hundred points for combined productivity and journal prestige; men, only twenty, a difference giving men a five-to-one advantage over women.[12] And, as Harriet Zuckerman, a sociologist of science, explains, advantage accumulates: ". . . When certain groups receive greater opportunities to enlarge their contributions to knowledge and then are rewarded in accordance with those contributions . . . recipients are enriched at an accelerating rate and conversely non-recipients become relatively impoverished."[13]

SELF-CONCEPT

"Physics is hard," says one respondent. "And at every level women tend to value themselves less than men do." Whether this is a consequence of too few women in physics or of less aggressive behavior natural to women, only psychiatrists can tell us. But low self-evaluation doesn't seem to deter women in other fields as much as it does in physics. From a twenty-nine-year-old Ph.D. in physics, now working in industry, comes this insight: "Where the quality of your thought is constantly being tested against tough questioning, being an egotist helps." In college, she found her male counterparts to be excessively egotistical, even more than they were antifemale. Later she acknowledged that their overconfidence contributed to their success.

Related to some women's lack of confidence is the terrible sense imparted to young women, particularly in high school, that they have to choose "either to be a girl or to do physics." This is in large measure because physics remains an "elective" in U.S. high schools, intended only for "nerds." In high school, one young woman unhesitatingly chose physics, but when queried by a dean at her college, she said she wasn't sure she would be able to make the same choice in college. Another young woman remembers being deterred from going into physics because it is a field "that is so inaccessible to others, it prevents your friends and family from having any idea of what you do." Women, perhaps more than men, want to be able to share with their intimates what it is they love and do.

Since physicists employ fewer lab technicians than either chemists or biologists, there are even fewer women around to counter the view that physics is not for girls. Even if lab workers have been grossly underpaid and underutilized, female students leaning toward chemistry and/or the biosciences have had a population of women professionals working in those fields to whom they can relate.[14]

WORK-FAMILY ISSUES

Numerous studies have shown that married women scientists with children are just as productive as women scientists without. But that's an after-the-fact analysis. From the point of view of a young woman choosing physics, the possibility of combining physics and family looks grim. To quote a respondent:

> First, there is the biological clock. A typical female will spend seven years earning a Ph.D. making $16K a year with minimal health insurance and no retirement and then do four years as a postdoc. As a graduate student, she must take classes, teach undergraduates, and perform research. She's too busy to have kids. As a postdoc in her early thirties, she is in her most productive research years and she's got to make them count so she can look good for a tenure track position. Not a good time to have kids, either, not if she's a serious researcher. So at thirty-four, if she's headed toward an academic career, she can foresee seven more years of teaching and doing research. Having a kid during this time means she's not serious about physics.

Less widely reported is the "need to nurture" that women physicists carry over with them from home to their teaching. From a just-tenured professor of physics at a flagship university in the mid-South:

> The problem? Too many things to do. Women in academe get dumped on because we are uncomfortable doing a sloppy job at things that impact *people*. We are not as willing to justify a

hands-off attitude that, for example, something should be the students' responsibility and not yours. So you get used up, and the more successful you are, the more stressed. Meanwhile, your female students, who look to you as a role model probably say "No way. I want a nine-to-five job that pays well and leaves me time for my family, and I can get that with a B.S. in Physics."

This may be why, as a senior woman physicist at an Ivy League institution reports, "I see lots of women in my advanced introductory physics class—sharp, eager, bright young women. But among my women seniors I hear doubts expressed about pursuing a straight research career. There's talk of taking a year off or pursuing physics education or public policy rather than straight physics."

Perhaps it's the absence of discussion of their family ties among male physicists on the job, or the even more worrisome forbiddance of mention of a deceased physicist's family or outside interests in any obituary published in *Physics Today* that conveys that family is, essentially, irrelevant to "the work."

Everything said by my respondents about work-life conflicts is true of academe in general. Yet, thousands of women in other fields get through, despite the demands of the academic life. In the humanities, although the ultimate earnings are less than in physics, at least the apprenticeship (without postdoc) is two to four years shorter. But otherwise, the need to do research, produce, excel at teaching, take on committee work—just when you're busiest—is the same as in physics, and women do it. Women in bioscience do it, too. So there must be something about the physics research career that stymies women. Could it be the belief, right or wrong, that the best work is done before age thirty, so the competition to show brilliance is intense (for mathematics even more so)? In the field in which I trained, we get better as we get older, and everybody knows this. In physics, even where that is equally true, it's not widely believed.

In comparison with academe, one respondent, working in aerospace, says industry offers much better opportunities for women in physics. "Industry provides both technical and management ladders, which allow women to choose career options

depending upon their personalities and skills. Industry also provides more benefits which are women friendly: health insurance,
401K retirement, tuition, maternity leave, matching day care
funds, high salary. Overall, industry provides both the environment and the benefits that allow a female with a degree in physics
to succeed." But industry is rarely presented as an option in
physics. Although nearly 70 percent of the 1997–1998 Ph.D. graduates in physics got what the American Institute of Physics calls
"potentially permanent positions" in industry, the physics major
is still seen to be a ticket to graduate school, and graduate school
a ticket to the professoriate. So the constraints of academe are
seen by the aspiring female physicist to be what she will have to
contend with.

MAKING CHANGE

"What to do?" a senior woman physicist working in a national lab
echoes my question. "Look at engineering, where the numbers of
professional women were just as small as in physics two decades
ago but where there has been leadership and directives from the
top. A number of Deans of Engineering just decided they were
going to recruit women students and hire women faculty, and they
are doing so." Twenty percent of those employed at the
instructor/lectureship level in engineering throughout the country
are women, which in time should change the ratio of full professors (now just 2.2 percent). Physics functions at the level of
departments, with rotating chairs who don't have much power or
resources. Engineering deans have both. Also, they are connected
to industry, which, in an environment dominated by equality in
employment issues has had to be more forward-looking than
academe as regards women.

Indeed, a physicist working in an engineering department at a
large state university reports that her students are not at all surprised to see her at the chalkboard because in their first-year
required design course two out of five instructors are women.

Certain physicists are making change. My respondents have

bouquets for individual members of the profession who are outproducing women Ph.D.s, in contrast to the country's "Bluebeards," so named because women never emerge from their doctoral programs, at all or intact. Certain undergraduate programs are developing unusually large proportions of women students. At Dickinson College, where there is both Priscilla Laws and Workshop Physics, for the past ten years 40 percent of the physics majors have been women. Reports Laws, "Of those who take the calculus based Workshop Physics, women are just as likely as men to major in physics. Thirty-six percent of the awards for outstanding performance in workshop physics over the past fourteen years have gone to women." However, even at Dickinson, women rate their own mastery of physics a full point below that of their male counterparts in spite of the fact that their average GPA is the same. This is consistent with other studies showing women to be less confident.

Some respondents are looking ahead to a changed environment as " '90s males" move into power. Another is more skeptical: "It's not enough to wait for the Old Guard to die. Don't forget; it is they who are training the New Guard."

THE WAY IT'S 'SPOSED TO BE

Debora Katz, associate professor of physics at the U.S Naval Academy writes:

> I love physics. I fell in love when I was in high school physics in Hawaii. . . . In high school, I won the chemistry prize and my chemistry teacher strongly encouraged me to pursue a career in chemistry. When I took computer science, my teacher encouraged me to pursue a career in computer science, and when I was in physics, my teacher encouraged me to pursue a career in physics. When I went to college, I figured I would be pre-med (that made the parents happy). . . . I took a mathematical methods for physics course. I quickly learned that I didn't like being pre-med. I was not wild about chemistry, but I really enjoyed physics. [In a pre-med chemistry class, when she scored

the highest grade in the first lab exam, her lab station was sabo-
taged by other pre-meds.]

. . . My advanced level physics class was small. Most of the students had had a lot more physics and math than I had. So I met
with the professor every week for private tutorials. He was a terrific teacher . . . and the physics department was a supportive
place. I majored in physics and dropped the pre-med program. . . .

I learned that I was a freak of nature only when the same
professor who had encouraged me to work with him during the
summer was helping me apply to graduate school. He asked why
I wanted to go into experimental physics; he said it was unusual
for women. He wanted to know if my father had given me ham-
mers to play with as a child. I had no answer. I just liked physics.
Do I have to come up with some reason that my male counterparts don't need?

In graduate school, I learned that physics is hard. I had no
idea. It came pretty naturally to me as an undergraduate, but of
course graduate work is harder. I was the only woman in my
class of sixteen students. I made some of my best friends in graduate school. Without collaborating with them, I would have
probably dropped out. I feel that graduate school still has its
hazing aspects. . . . I had won several fellowships in graduate
school. I remember a minority of male students thought I got my
fellowships because I was a woman. . . . I knew I worked hard
and had terrific grades and recommendations.

As a professor . . . I am very fortunate. I am in a department
with four other female professors [out of thirty-five]. They are
professional, supportive and friendly. For the first time, I feel
that I can be feminine, professional and well respected.

Amen.[15]

NOTES

1. Rachel Ivie and Katie Stowe, eds., *Women in Physics* (College
Park, Md.: American Institute of Physics, 2000), highlights.
2. U.S. physicists are not alone. The low participation rate of U.S.
women in physics is the same as that of Europe's Protestant countries.

In Catholic and Orthodox Europe, where high schools are still sex-seg-regated and physics is often a required course for university-bound students, the pattern is different.

3. See the works of David F. Noble, *A World without Women: The Christian Clerical Culture of Western Science* (New York: Oxford University Press, 1992) and Margaret Wertheim, *Pythagoras' Trousers: God, Physics, and the Gender Wars* (New York: W. W. Norton, 1995).

4. Vivian Gornick interviewed the then eighty-four-year-old I. I. Rabi in 1982. He, who had never had a woman graduate student, told Gornick that women were "unsuited for science." It was, as he explained to her, a matter of the nervous system. "It makes it impossible for them to stay with the thing. . . . Women may go into science . . . but they will never do Great Science." Vivian Gornick, *Women in Science* (New York: Simon and Schuster, 1983), p. 36.

5. Gerhard Sonnert and Gerald Holton, *Gender Differences in Science Careers* (New Brunswick, N.J.: Rutgers University Press, 1995).

6. Rita Colwell, Introduction to Elga R. Wasserman, *The Door in the Dream: Conversations with Eminent Women in Science* (Washington, D.C.: Joseph Henry Press, National Academy of Sciences, 2000). In Sonnert's and Holton's study, cited above, for all the sciences, 75 percent of women experienced gender discrimination; 12 percent of men did.

7. Laurie McNeil and Marc Sher, "The Dual-Career-Couple Problem," *Physics Today* 32–37 (July 1999), chart, p. 3., based on an APS Membership Survey that also documents that of the 74 percent of male physicists who are married, 82 percent are not married to women scientists.

8. Nancy Hopkins, *A Study of Women Faculty in Science at MIT* (Cambridge, Mass.: MIT, 1999).

9. Wasserman, *The Door in the Dream*, p. 200.

10. Ibid., p. 88.

11. Ibid., p. 118.

12. Ibid., p. 183.

13. Harriet Zuckerman, Jonathan R. Cole, John T. Bruer, eds., *The Outer Circle: Women in the Scientific Community* (New York: W. W. Norton, 1991), p. 53.

14. This insight comes from a personal communication from Elga Wasserman.

15. The author wishes to express her appreciation to Dr. Eleanor Babko of the Council of Professionals in Science and Technology for providing so much of the data quoted in this article and much guidance overall.

Status of the "Two Cultures"

Melvin Schwartz

Some forty years ago C. P. Snow wrote a book titled *The Two Cultures* in which he examined the increasing separation between the scientific culture and the nonscientific culture.[1] It was not always so. Back in the days of Newton and Maxwell, science was called *natural philosophy*, and the truly literate person was expected to be as versed in the motion of the planets as in Shakespeare. Unfortunately, that is no longer the case. It is the rare scientist who can talk about literature and the arts; on the other hand, there aren't many writers and artists who understand relativity and quantum mechanics. We ask whether this situation must remain so or if by changing the way we educate our young people we can come back to the point where the two cultures are once again merged in the minds of literate people.

Some ten years ago I chaired a committee to plan the future of Columbia College (Columbia College is the major undergraduate institution in the Columbia University system), in particular with respect to its core curriculum. For many years Columbia has required that each of its students take a two-year sequence of courses covering literature, history, philosophy, music, and art. In fact, one of the courses I enjoyed most as a Columbia student was Literature

Humanities. I happened to be taught by a renowned Dante scholar, Professor Joseph Mazzeo, and the course gave me an insight into the world beyond science. Very few scientists, particularly those educated at engineering schools, have any feeling for the intellectual content beyond the sciences.

On the other hand, the core curriculum has a minimal program for the nonscientist. Rather than insist that every student know the basics of modern science and mathematics, the college accepts a conglomeration of relatively trivial and uninteresting courses to satisfy the so-called science requirement. Thus we have a situation where a student can graduate from Columbia College without knowing what quantum mechanics is or how to use calculus. No matter that calculus is really cunning addition and subtraction; very few nonscientific intellectuals don't shudder with fear when the word is spoken. I proposed to my committee that science should be treated just like humanities; each and every student should be required to take a two-year sequence wherein the major discoveries in physics, biology, chemistry, and mathematics would be covered.

Most of the other scientists on my committee were completely opposed to the idea. They felt that teaching freshmen would detract from their research time and since their careers depended on the number of papers they could turn out, there was no incentive for participating in such a program. Unless we can have a separate teaching faculty that derives recognition and tenure from interaction with undergraduates, the notion of making science into an intellectual activity enjoyed by nonscientists is just a pipe dream.

Let me say a word or two about mathematics. To most people mathematics is just like arithmetic; we must learn it in order to function in the world of technology. At the beginning we learn how to add, subtract, multiply, and divide. Then we learn calculus in order to be technologically viable. At no point in the nonscientist's education is he likely to come across the important concepts of mathematics such as group theory, number theory, complex analysis, topology, and differential geometry. Real mathematics is much like poetry; it represents in many ways the highest point in human thought. And if we are were looking for role models, it

would be good to remember that one of the great minds of twentieth-century philosophy was Bertrand Russell.

Much of the split between the two cultures, scientific and nonscientific, arises because people fail to distinguish between science and technology. When you mention quantum mechanics,
people think of atomic bombs. When you mention electromag-
netic theory, most people think of television sets. It's like thinking
of reading cookbooks when you mention Shakespeare. If you don't
care how television sets work, should you really care about electromagnetism?

This confusion of science with technology led to the absurdity
of a U.S. senator demanding that the National Science Foundation
fund only such research as is relevant to the nation's needs. I am
reminded of the time an interviewer asked Michael Faraday if
magnetic induction had any practical value. He replied that he
didn't know, but he was sure the government would find some way
to tax it. You can be sure that Isaac Newton did not search for
immediate relevance when he set out the laws of planetary
motion. Nor did J. Clerk Maxwell search for relevance when he set
forth the basic equations of electromagnetism.

This century has had two great revolutions in our understanding of the physical world around us. The first of these—relativity—united electricity and magnetism and showed how
Maxwell's Equations were an essential outcome of Coulomb's Law,
wherein two charges repel or attract one another with a force proportional to the inverse square of the distance between them. The
other great intellectual revolution was quantum mechanics, which
explained the behavior of atoms and molecules on the microscopic level.

Even though these great discoveries were made without
thought as to the practical consequences, they did have many
such uses. Indeed, much of what makes our lives comfortable is
the result of discovering the applications of pure science. An
understanding of the behavior of doped silicon led to integrated
circuits and desktop computers. An understanding of the structure
of large molecules led to the discovery of DNA. Molecular biology,
which didn't exist when I was in high school, now promises to cure

many of our diseases and perhaps even make us immortal some day. When you start out with the desire of understanding the basic laws of nature, who knows where it will lead you?

The rift between the two cultures is not entirely the fault of nonscientists. We must convince many of them that there is much more to enjoy beyond the strictly positivist world of the scientist. There should be room for art, music, history, philosophy, and literature in the scientist's life. Perhaps in this way we can remove the rift and reunify the two cultures.

NOTE

1. C. P. Snow, *The Two Cultures and the Scientific Revolution* (Cambridge, Mass.: Cambridge University Press, 1960; reissue edition 1993).

IGNITING AN EDUCATIONAL REVOLUTION

MIT OpenCourseWare

Charles M. Vest

At the Massachusetts Institute of Technology (MIT), we see the quality of our shared human future on earth as a problem to be solved, a problem massively difficult, but massively worth the solving, and a great deal of our scholarship and research is dedicated to that end. At the same time, MIT strives to lead the world not only in delivering but in designing an exceptionally rigorous science-based education for exceptional students from around the globe. This spring, those two interests—in world-changing research and world-class teaching—gave birth to a single project we hope may be as significant as any scientific or technological advance ever to emerge from our labs: MIT OpenCourseWare.

On April 4, 2001, MIT announced that it would make virtually all of its undergraduate and graduate course materials available, free of charge, to anyone, anywhere on Earth, through the World Wide Web. At the initial press conference, one reporter asked, in effect, Why in the world would MIT professors give away their ideas for free? Wouldn't they be concerned about the lost potential for income? A faculty member on hand gave the simple reply: For scholars and teachers, the greatest possible reward is the sense that other people find their ideas important, inspiring, useful—that their work can indeed change the world.

In a market-driven universe, where the Internet and the Web are increasingly seen as vehicles for making money, MIT Open-CourseWare seems counterintuitive. But as any scientist can tell you, the most exciting new ideas usually do.

OpenCourseWare emerged as the surprising result of an intensive faculty and staff effort to decide what major initiative MIT should undertake in the evolving field of distance learning. Eventually, we came to see the great pedagogical challenge of the moment as defining what the Internet revolution will mean, and should mean, for higher learning on this planet.

We are faced, of course, with the stark fact that even now only 5 percent of the world's population has access to computers and the Internet. Obviously, to make the conversation relevant for the vast majority of humanity, a critical priority must be to provide greater access for many more people around the globe. Working together, government, industry, academia, and nongovernmental organizations must find the will and the hardware to bridge the digital divide.

But frankly, this technical challenge is the easy part—extraordinarily expensive, perhaps, but easy. For us at MIT, the real issue was the intellectual one. As increasing numbers of people and institutions gain online access, how should all that technology best be used for education? And especially, how should it be used to make the most difference for those in greatest need—those parts of the world isolated by geography, poverty, or politics?

One vision, popular since the birth of the Internet, is a kind of Universal-College-from-Your-Keyboard: Take the best professors in the world in each subject, capture their lectures electronically, and make them available worldwide.

That idea still has advocates, but it has its limitations. Taken to the extreme, hundreds of thousands of students worldwide receiving precisely the same lecture from the same professor on a machine is a nightmare in my view. Furthermore, it mirrors the model of *business-to-consumer*, or B2C, electronic commerce. B2C electronic commerce is somewhat interesting and important. But it is business-to-business, or B2B, electronic commerce that has been the truly transforming influence of the Internet in business and industry—and we believe the same will be true in education. The direct merchandizing of university courses, even interactive ones, will definitely have a role in global education. But the real power will arise as faculty in colleges and universities all over the world openly share educational materials with each other.

It is that intellectual revolution we hope MIT OpenCourseWare will ignite. The material made available will include detailed lecture notes, course outlines, reading lists, problem sets, essay topics, simulations, and demonstrations. Having access to those resources will *not* mean that users will be able to earn an MIT degree online, or even academic credit; OpenCourseWare is not an attempt at interactive distance learning. It *will* mean, however, that educators and students around the world will be able to select whatever pieces interest them, add their own flavor, and shape them for use in the context of their own research, curriculum, culture, and goals. It will mean that the cumulative wisdom of our faculty—not just as scholars but as expert teachers—will be available to help other educators instruct and inspire their own students. And it will mean that as new knowledge and educational content emerge, we can disseminate them around the world instantly—a key step toward closing the gap between the information "haves" and "have-nots."

It is no secret that university faculty have always shared such materials informally with a few colleagues and former students at similar institutions. Now we must do this in a globally open manner and with the speed of Internet time. Together, the creators and users of these materials will weave a new worldwide web of knowledge and learning that will complement and stimulate innovation in ways we cannot even imagine today.

We certainly hope MIT OpenCourseWare will be of use to individuals—from the precocious high school senior studying biology in Singapore to the city planner battling sprawl in Madrid and even to the MIT student in bed with the flu and unable to make it to class! But the truth is that most learning, especially within colleges and universities, will remain a deeply human activity, built on an ancient model of mentorship and dialogue. Information technology will enhance the role of the teacher, but it will never supplant it. And anyone who has ever been to college can attest that what they learned, they learned in part—perhaps in large part—from their fellow students. Those are dimensions no amount of information alone can replace.

So again, following the B2B and open system models, our true audience is not primarily individual students, but rather our global colleagues in education—from faculty members launching a new engineering university in Ghana to a professor in Rio seeking better ways to convey the deep mysteries of economics. The real power of OpenCourseWare will come in the ways it allows us to share our strength with other faculties and institutions.

Our excitement about the potential of OpenCourseWare has been reflected in the support we have received worldwide. Since the announcement in April 2001, we have received literally thousands of messages of support and enthusiasm from around the globe.

Especially encouraging, of course, has been the indispensable and generous support of the Andrew W. Mellon Foundation and the William and Flora Hewlett Foundation, who together will fund the project's crucial start-up and pilot phase, scheduled to begin in the fall of 2002.[1] Experience gained from the first phase will help determine the costs and challenges of the second phase, expected to take an additional six years.

With the initiative now seriously under way, we are naturally looking to the future, and to our final great vision for the project, that other institutions will be willing and able to throw open their intellectual doorways as well—that OpenCourseWare will be a beautifully contagious idea. "Our hope," said Paul Brest, president of the Hewlett Foundation, "is that this project will inspire similar efforts at other institutions and will reinforce the concept that ideas are best viewed as the common property of all of us, not as proprietary products intended to generate profits."

MIT OpenCourseWare may go against the grain of today's dominant market values, but it is a great celebration of the democratizing openness and opportunity inherent in the Internet and the Web. And it is a perfect expression of the core values and traditions of MIT and of all the world's finest institutions of higher learning. OpenCourseWare is built on the belief that the most powerful, transformative force on Earth may well be education; that the soul of education and human development is the free and open sharing of information, philosophy, and modes of thought; and that the permanent challenge of educators is to widen the world's access to information and ideas and encourage others to do the same. As we open our own electronic doorways, we look out on the dawning with tremendous hope.

NOTE

1. Since this essay was written, MIT OpenCourseWare has launched its first offerings on the Web—a pilot giving thirty representatives courses from all five of the institute's schools. During its first month of operation, the site (http://ocw.mit.edu) received 42 million hits from 315,000 unique visitors, at least 30 percent of them from outside the United States. Visitors have come from 177 different countries and all seven continents, including Antarctica. The institute has received well over three thousand e-mails about the site, almost unanimously grateful or congratulatory. This response has exceeded our highest expectations and confirms our belief that OpenCourseWare has the potential to help transform education around the world.

PART 2
REFRAMING SCIENCE LEARNING

ON CREATING A "SCIENTIFIC TEMPER"

Bruce Alberts

It is a pleasure to have the honor of recognizing the many contributions of Leon Lederman to education by means of this brief essay on the occasion of his eightieth birthday. Whenever I hear Leon produce one of his fascinating riffs on a current aspect of science that excites him, I experience a flashback—remembering how exciting and intriguing the many mysteries of the world were to me, as I encountered them during my childhood. I am also reminded of the excitement for learning that one can witness when visiting any kindergarten class—as the five-year-olds explore the many wonders that they find in their surroundings.

What should be the central goal of education for life in the twenty-first century, in a society that will experience many rapid changes due to the continuing startling advances produced by science and technology? I claim that the goal must be to maintain the zest for learning of young children—as they progress through thirteen or more years of our formal education systems—while at the same time providing them with the skills that they will need to become effective lifelong learners. Anyone who visits a typical eighth-grade class will recognize that we are far from this goal today. For a vast number of those thirteen-year-olds, school has become a nearly irrelevant exercise, seemingly unconnected to their lives out of school and unable to compete with the many distractions of our mass media–saturated society. And perhaps most frustrating to Lederman, by the time that these students graduate from high school, they will rank at the very bottom in international comparisons of mathematics and science comprehension.

How can the world's most scientifically and technologically advanced nation produce, year after year, seventeen-year-olds with such poor achievements in science and math? At a fundamental level, the answer is simple. For the most part, the large, vibrant science and engineering communities in the United States have been completely disconnected from the science, technology, and math education provided to students in the precollege years. Worse than that, the science and math departments in our universities have generally failed to teach their introductory courses in ways that excite students, enable them to directly experience scientific problem-solving through exposure to inquiry-based learning, and clearly connect their education to the world they are experiencing outside of the formal learning environment. Today's science, engineering, and mathematics faculties are thereby defining their discipline for our future parents and teachers in ways that conflict with the type of science, mathematics, and technology learning that have been forcefully presented in the recent national standards documents in these three disciplines (National Research Council 1996; American Association for the Advancement of Science 1993; National Council of Teachers of Mathematics 2000; Technology for All Americas Project 2000). For example, the standards encourage high school teachers to teach an introductory biology course as a series of in-depth explorations in which students come to understand a few central topics deeply. But it is foolish to expect this to happen if we define Biology 101 in our prestigious universities differently: as the Herculean task of sampling all of the knowledge that has been discovered by biologists about the living world so as to be able to repeat it back on exams.

What is it that we want to achieve through science education in our schools and colleges? Most scientists will answer that we want to discover and encourage those rare students who have the ability, temperament, and interest to become the leading scientists of the next generation. Because the professors who are scientists were themselves selected by a system that taught biology, chemistry, and physics as large lecture courses with cookbook laboratories—and no inquiry—their natural tendency is to assume that the education system that they experienced represents the "best of all possible worlds." They were, after all, the A-plus students in these courses, automatically selected as those who learned well in this standard way. But we now know that different students learn best in different ways, and it is clear that our introductory courses are quickly eliminating large numbers of people who could have become outstanding scientists if only we had provided them with more than one way to discover their talents and interests.

In a world increasingly dominated by science and technology, one might have hoped that most science professors would give a broader answer to the question I posed above. Of course, we want to encourage future scientists, but we also want to achieve a much broader goal, one that Jawaharlal Nehru wished for India half a century ago: creating—through our science courses—a "scientific temper" for our nation (National Research Council 1998).

It has been frustrating for both Lederman and me to observe the glacial pace at which we have been able to improve the quality of the science education that we provide to all students. After all, the stars would seem to all be in alignment. As witnessed by the detailed surveys and analyses in the book *Teaching the New Basic Skills* by Richard Murnane and Frank Levy, the business community is seemingly unanimous in its strong desire for high school graduates who can solve problems, think quantitatively, and, more generally, "think for a living" (Murnane and Levy 1996). Reading their descriptions of the deficiencies of the U.S. workforce, one cannot help but be struck by the seemingly perfect match between creating a high school graduate capable of being a productive contributor to our economy and the type of inquiry-based science education recommended for all students ages five to eighteen in the *National Science Education Standards* (see http://www.nap.edu/catalog/9596.html). And precisely the same type of logical, problem-solving skills are badly needed by all citizens, if they are to make wise choices when confronted with the enormous number of personal, community, and national decisions that they will need to make in our ever more complex democracy.

Bob Galvin, a leader of modern industry who has had to retrain large numbers of poorly prepared high school graduates so that they can function in entry-level jobs at Motorola, says:

> While most descriptions of necessary skills for children do not list "learning to learn," this should be the capstone skill upon which all others depend. Memorized facts, which are the basis for most testing done in schools today, are of little use in the age in which information is doubling every two or three years. We have expert systems in computers and the Internet that can provide the facts we need when we need them. Our work force needs to utilize facts to assist in developing solutions to problems.

What will it take to make real progress in providing a quality education for all Americans? First of all, it will take a permanent commitment by the professional classes to connect directly to the world of the K–12 teacher. To listen to these teachers' voices, their hopes, and their ideas so as to be effective advocates and supporters for a vastly improved public education system—starting in their local schools. I have spent more than eight years as president of the National Academy of Sciences, where one of my major goals has been to improve the quality of science education in our nation's schools. I believe that one of the most important tasks now facing the academy is catalyzing the creation of a productive two-way connection between our nation's best scientists and the many heroic and dedicated teachers of science at all levels. In this critical effort, Leon Lederman has set an inspiring example for us all.

In summary, we all need to remember that the challenge for those who want to improve education is to create an educational system that exploits the natural curiosity of children, so that they maintain their motivation for learning not only during their school years—but throughout life.

I end with a quote from Richard Feynman, who like Lederman, was both a distinguished physicist and an inspiring communicator. One summer, in the Catskill Mountains of New York when Feynman was a boy, another boy asked him, "See that bird? What kind of bird is that?" Feynman answered, "I haven't the slightest idea." The other boy replied, "Your father doesn't teach you anything!" But Feynman's father *had* taught Feynman about the bird—though in his own way. As Feynman recalls his father's words:

> See that bird? It's a Spencer's warbler. (I knew he didn't know the real name.) . . . You can know the name of that bird in all the languages of the world, but when you're finished, you'll know absolutely nothing whatever about the bird. You'll only know about humans in different places and what they call the bird. So let's look at the bird and see what it's doing—that's what counts.

REFERENCES

American Association for the Advancement of Science. *Benchmarks for Science Literacy*. New York: Oxford University Press, 1993.

Murnane, Richard J., and Frank Levy. *Teaching the New Basic Skills: Principles for Educating Children to Thrive in a Changing Economy*. New York: Simon and Schuster, Inc., 1996.

National Council of Teachers of Mathematics. *Principles and Standards for School Mathematics*. Reston, Va.: National Council of Teachers of Mathematics, 2000.

National Research Council. *Every Child a Scientist: Achieving Scientific Literacy for All*. Washington, D.C.: National Academy Press, 1998.

National Research Council. *National Science Education Standards*. Washington, D.C.: National Academy Press, 1996.

Technology for All Americans Project. *Standards for Technological Literacy: Content for the Study of Technology*. Reston, Va.: International Technology Education Association, 2000.

RETHINKING THE PHYSICAL SCIENCES IN SCHOOL PROGRAMS

Rodger W. Bybee

This essay, written to honor Leon Lederman's contributions to science education, uses the physical sciences as a specific example for a broader discussion of curriculum reform. I provide some insights about the *National Science Education Standards* (National Research Council, hereafter NRC 1996) and reports from the *Third International Mathematics and Science Study* (TIMSS) (U.S. Department of Education 1996, 1997, 1998). Through this discussion I hope to expand our perspective of standards-based reform in science education and provide insights about designing and delivering science curricula in the twenty-first century.

WHY STANDARDS-BASED CURRICULUM?

The power of standards lies in their capacity to change fundamental components of the educational system. This premise has several key points worth noting. Standards have the capacity to cause or influence change, but the actual changes will vary as the standards are interpreted in terms of curriculum programs, instructional practices, and educational policies designed to implement and sustain the changes implied by the standards. The standards imply changes at the instructional core, by which I mean curriculum content, instructional techniques, assessment strategies, teacher education, and professional development programs. This point is important in a reform era where we see numerous innovations that may have merit but are not necessarily directed toward fundamental changes that have a high potential of enhancing student learning. I am referring to innovations such as site-based management, vouchers, and charter schools. To be clear, the evaluation of such innovations ultimately lies in student learning, not their political value.

Standards influence the entire educational system by nature of the fact that they are input, but they also define *output* for which we use the defining question "What should all students know and be able to do?" In educational history we have primarily focused on *inputs* with the hope of improving *outputs*—greater student learning. So, for example, we change the length of the school year, science courses, textbooks, technologies, and teaching techniques. All such inputs are meant to enhance learning, but they have been inconsistent, not directed toward a common purpose, and centered on different aspects of the educational system. A lack of coherence and consistency characterizes many contemporary analyses of the science curriculum.

NATIONAL STANDARDS AND THE PHYSICAL SCIENCES

Before discussing the national standards, the reader should understand the perspective that guided our work on that project. For the majority of students, tenth-grade biology is their final experience with science in the K–12 school curriculum. Stated another way, because chemistry and physics are generally taught at eleventh and twelfth grades, respectively, a majority of students do not take science courses that introduce physical science concepts basic to understanding the natural world. One goal of the national standards was to present the science content and recommendations in such a manner that would result in the physical sciences becoming a part of students' experiences in high school.

To be clear and direct, the national standards recommend that *all* students develop and understand fundamental concepts associated with physical sciences.

Although most scientists and science educators recognize the categories in Table 1 and affirm the standards (they almost always suggest other science concepts or recommend elaborations of those in the standards, but this is another issue that deserves discussion), they often miss the educational and political realities of the recommendation. The intrepid nature of the recommendation lies in the fact that it applies to *all* students and thus implies that something should change in the way schools organize the science program, especially the high school science curriculum. Namely, the *National Science Education Standards* (NRC 1996) has set in place policies that, if implemented, allow all students the opportunity to learn fundamental concepts in the physical sciences. For those states, schools, and science teachers paying attention, curricular change becomes an inevitable consequence of the national standards.

One should note a subtle but essential point: The national standards broadly define the content of physical science and not the school curriculum. I quote from the national standards:

> Curriculum is the way content is delivered; it includes the structure, organization, balance, and presentation of the content in the classroom. . . . The content standards are not science lessons, classes, courses of study, or school science programs. The components of science content described can be organized with a variety of emphases and perspectives into many different curricula. The organizational schemes of the content standards are not intended to be used as curricula; instead, the scope, sequence, and coordination of concepts, processes, and topics are left to those who design and implement curricula in science programs. (U.S. Department of Education 1996)

Clearly, the national standards leave curricular structure to professional curriculum developers, such as Biological Sciences Curriculum Study (BSCS); to states and school systems, such as California and San Diego Unified School District; and to scientists and science education leaders such as Leon Lederman.

TABLE 1. BSCS SCIENCE: AN INQUIRY APPROACH

Grades 6–11 Framework

	6	7	8	9	10	11
	Let's Take a Look				Let's Take a Closer Look	
Science as inquiry (2 weeks)	Science as a way of knowing	Science as a way of knowing	Science as a way of knowing	Science as a way of knowing	Science as a way of knowing	Science as a way of knowing
Core concepts (physical) (8 weeks)	• Properties and changes in properties of matter • Integrating chapter	• Motions and forces • Integrating chapter	• Transfer of energy • Integrating chapter	• Structure of atoms • Structure and properties of matter • Integrating chapter	• Chemical reactions • Motions and forces • Integrating chapter	• Interactions of energy and matter • Conservation of energy and increase in disorder • Integrating chapter
Core concepts (life) (8 weeks)	• Structure and function in living systems • Reproduction and heredity • Integrating chapter	• Diversity and adaptations of organisms • Populations and ecosystems • Integrating chapter	• Regulation and behavior • Integrating chapter	• The cell • Behavior of organisms • Integrating chapter	• Molecular basis of heredity • Biological evolution • Integrating chapter	• Matter, energy, and organization in living systems • Interdependence of organisms • Integrating chapter
Core concepts (earth-space) (8 weeks)	• Structure of Earth systems • Integrating chapter	• Earth's history • Integrating chapter	• Earth in the solar system • Integrating chapter	• Geochemical cycles • Integrating chapter	• Origin and evolution of the Earth system • Origin and evolution of the universe • Integrating chapter	• Energy in the Earth system • Integrating chapter
History and nature of science (2 weeks)	• Science as a human endeavor	• Nature of science	• History of science	• Science as a human endeavor	• Nature of scientific knowledge	• Historical perspective
Science in a personal and social perspective/ Science and Technology (8 weeks)	• Natural hazards • Risks and benefits • Abilities of technological design	• Personal health • Populations, resources, and environments	• Science and technology in society • Understandings about science and technology	• Personal and community health • Natural and human-induced hazards • Abilities of technological design • Environmental quality	• Population growth • Natural resources • Environmental quality • Personal & community health	• Science and technology in local, national, and global challenges • Understandings about science and technology • Natural resources

Standards-based modules

Last revision: 10/10/02

———— In development

TIMSS AND THE PHYSICAL SCIENCES

Results from the Third International Mathematics and Science Study seized educators' and the public's attention in the late 1990s, the same period as national standards. Combined, the poor achievement on TIMSS and new standards for science proved to be a powerful force to rethink the science curriculum, especially in the physical sciences.

TIMSS provided a scenario of student achievement from the early grades through high school. The results portrayed disappointing insights; U.S. students demonstrated decreasing achievement with increasing grades. When compared with other countries, the more time in school, the lower student achievement in science. The following statements from the National Center for Education Statistics (NCES) reports on TIMSS summarize the scenario of student achievement for fourth-, eighth-, and twelfth-grade students:

- U.S. fourth-graders score above average in both science and mathematics compared with the twenty-six nations in the TIMSS fourth grade assessment. . . . In science, 16 percent of U.S. fourth-graders would rank among the world's top 10 percent. . . . our students are outperformed by only one country—Korea (U.S. Department of Education 1997).
- U.S. eighth-graders score below average in mathematics achievement and above average in science achievement, compared to the forty-one nations in the TIMSS assessment. In science, our eighth-graders' international standing is stronger in earth science, life science, and environmental issues than in chemistry and physics (U.S. Department of Education 1996).
- U.S. twelfth-graders scored below the international average and among the lowest of the twenty-one TIMSS nations in both mathematics and science general knowledge in the final year of secondary school (U.S. Department of Education 1998).

In light of this particular discussion, fourth-graders score lower (but above average) in the physical sciences compared to the other sciences (U.S. Department of Education 1997). By eighth grade, students' scores in physics and chemistry are not significantly different from the international average, but they are lower than U.S. scores for earth science, life science, and environmental issues. In fact, our better-than-average scores on environmental issues, earth science, and life science account for the overall science score being above average (U.S. Department of Education 1996). By twelfth grade, U.S. students' scores are among the lowest, and achievement in the physical sciences is lowest of scores for all the sciences.

The aforementioned statements describe the achievement of *all* students. Even more disappointing was the achievement of advanced students. To quote, "The performance of U.S. physics and advanced mathematics students was among the lowest of the sixteen countries that administered the physics and advanced mathematics assessments (U.S. Department of Education 1998). In fact, no countries scored below the United States on the physics assessment. U.S. students who had taken or were taking advanced placement physics did better, they scored better than France and the Czech Republic and significantly higher than Austria."

The conclusion I draw from these results is that it is time to rethink the science curriculum, especially in the physical sciences, and the *National Science Education Standards* (NRC 1996) can be used as the basis for the physical sciences. In rethinking the science curriculum, we must address several fundamental curricular issues.

REDESIGNING THE SCIENCE CURRICULUM

What students learn in school results from many factors. This insight seems relatively clear. One obvious factor is the curriculum and another is the opportunity students have to learn valued content. TIMSS provides examples of countries that performed better and worse on content emphasized in the respective countries. For example, in the United States, eighth-graders scored second among TIMSS countries on "life cycles and genetics," topics widely taught in elementary and middle schools. On "physical changes" our students scored near the bottom of TIMSS countries, reflecting the lower emphasis in U.S. curricula (NRC 1996; Schmidt and McKnight 1998).

One can consider diverse issues when redesigning the science curriculum. For instance, time, expectations, number of topics, order of topics, and the increasing/decreasing emphasis on topics. Relative to the amount of time devoted to science in fourth and eighth grades, U.S. students spend *more* time in science than many other TIMSS countries (U.S. Department of Education 1996). It seems our students have time to learn science, so there must be other features of the curriculum to consider as we rethink the place of physical science in the school program. Specifically, we can consider curricular focus, coherence, and academic rigor.

Focus in a curriculum measures the time, attention, and opportunity students have to learn content. U.S. students are introduced to more topics in less time than their peers in other countries. It is clear they learn less than their peers in TIMSS countries. This is an example of "more is less" when it comes to curricular focus and learning (Schmidt et al. 1999).

Coherence in a curriculum is another feature worth considering. Coherence measures the connectedness of science ideas and skills as students have opportunities to learn specific ideas and develop skills over time. In a coherent curriculum, content is introduced and developed smoothly and deliberately over time. Science content such as "properties of objects and materials" is introduced through activities in the K–4 grade level. These concepts then become the basis for more sophisticated activities on "properties and changes of properties of matter," and "structure of atoms" in grades 5–8 and 9–12.

Finally, I note a lack of rigor in science curricula. By rigor I am referring to the introduction and development of concepts and skills that are basic to science. Redesigning the science curriculum demands a clear, unified vision. Content must be focused, coherent, and rigorous.

NEW DESIGNS FOR PHYSICAL SCIENCE

If TIMSS provides a rationale for including more physical science in school programs and the national standards provide the content that should be taught, then we are left with curricular organization and problems of graduation requirements and the sequence of course offerings. One of the boldest contemporary recommendations comes from Leon Lederman and his suggestion to teach physics first in ninth grade. The physics course would be followed by chemistry and biology in tenth and eleventh grades, respectively (Lederman 2001; Bardeen and Lederman 1998). Such an approach works for those states and school districts that have a three-year science requirement for graduation. It certainly places physical science in a curricular position that would result in more students learning some basic science content such as energy transformations, forces and motion, and the structure and properties of matter.

One also could approach the new design for physical science by offering an integrated approach for grades nine, ten, and eleven. BSCS is currently working on the integrated program that has science content from the national standards. Table 1 displays the content for the BSCS program.

In either the physics first or the integrated approach, the new curricular design would require three years of science, be a significant change for school systems, and require professional development support for science teachers.

DELIVERING STANDARDS-BASED CURRICULUM: A COMMON ANSWER

What will it take to deliver physical science content in a standards-based curriculum? Until the release of the national standards, I would have answered this question in a fairly simple and straightforward manner. I would have provided a common educational answer. Use the standards for content as the foundation for designing, developing, and implementing instructional materials. In elaborating this recommendation, I would have described the characteristics of innovative instructional materials, the different instructional approaches to developing student understanding and abilities, the need to address concerns of science teachers, and the importance of administrative support for implementing new materials. I would have then presented the need to combine professional development with curriculum reform because ultimately, science teachers have the responsibility for establishing and developing the connections between the content of the curriculum and the students' scientific understanding and abilities.

The *National Science Education Standards* were released late in 1995, and we have all learned some important lessons that extend beyond the educational answer I just described. Although delivering a standards-based curriculum may be clear in educational theory, reform of the science curriculum is not that simple in a democratic society. Individuals and groups challenge the entire idea of standards and the specific content and orientation of national, state, and local standards. The political conflicts over standards has been more intense and acrimonious than many anticipated. In my view, we need to understand what happens as educational systems attempt to implement innovative science programs. Specifically, we need greater emphasis on and appreciation for civic discourse about important educational issues such as national standards and reform of the science curriculum. An appreciation for and application of civic discourse at national, state, and local levels could set the stage for delivering a standards-based curriculum. That said, there are other aspects of curriculum reform that we must recognize.

A BROADER PERSPECTIVE FOR CURRICULUM REFORM

Clearly, there is a need for exemplary instructional materials designed and developed to provide opportunities for students to learn the science content described in the standards. The discussion of issues related to physical science serves as one example of this point. So, too, there must be assessments aligned with that content. As I mentioned, professional development combined with curriculum reform is a fundamental feature of any contemporary answer to curriculum reform. This said, I would like to shift attention from an educational perspective that, to use an economic metaphor, focuses on supply to a perspective that seems a necessary and more sufficient response to issues of sustained curriculum reform. Staying with the economic metaphor, we should consider increasing a demand for standards-based programs.

I take the phrase used earlier, "delivering a standards-based curriculum," to mean much more than books, kits, or other materials arriving at the schoolhouse door. Delivering should literally mean that students experience a standards-based program. And to achieve this, I would first establish public support for the need to change the current science curriculum. Educators would have to make the case for standards and subsequent reform of science programs. Convincing the general public and changing state and local policies such as adoption requirements, resource allocation, and assessment practices seems basic to establishing a demand for standards-based curriculum. Developing a demand for standards-based programs establishes a countervailing force to the powerful marketing used to maintain traditional textbooks and tests in our schools.

Educators must respond to the question of incentives for initiating curriculum reform by those responsible for science education. What are the current incentives for school districts, schools, and science teachers to adopt a standards-based curriculum? I believe the answer is, there are none. We can argue the importance of "doing the right educational thing" for all students; but, in the face of state and local policies, budget allocations, parental pressures, administrative conflicts, peer criticisms, and other forces, the incentives for curriculum reform seem marginal at best. Economic, political, and many educational incentives serve to maintain the current approaches to school science programs.

One could reasonably ask about the role of high-stakes tests and their part in the current system of incentives for curriculum reform. Clearly, high-stakes assessments are affecting curricular reform, but they have not had the influence proposed in this essay. More times than not, we hear that science teachers have to divert from the curriculum and "teach to the test." To the degree that tests are an incentive for curricular reform, I would argue that they are regressive and not progressive in that they empha- size memorization of facts and lower levels of knowledge rather than higher levels of understanding. To the degree assessments are consistent with the standards, they represent an incentive for curricular reform.

Finally, I return to the need for professional development. This recommendation, however, has a slightly different cast from one that connects professional development and the implementa- tion of innovative instructional materials. I would argue the importance of educating science teachers about current under- standings of student learning, the complexities of designing stan- dards-based curriculum, and their role in curriculum reform. Pro- fessional development should promote change in the demand side of the curriculum reform equation. In brief, we should educate teachers and other professionals so they are clear about the need for, and qualities of, standards-based curriculum materials. At the end of the day, I believe, for example, textbooks would change due to market demands.

CONCLUSION

In this brief essay, the physical sciences identify several unique ideas that should be considered in any answer to questions that center on any rethinking of school science programs. I make no claim they are easy questions to answer or issues to resolve, only that they are important; they have seldom been addressed, and they must be part of any delivery of a standards-based cur- riculum.[1]

NOTE

1. Portions of this essay were presented at the Illinois Mathematics and Science Academy on April 12, 2000.

REFERENCES

Bardeen, Marjorie G., and Leon Lederman. "Coherence in Science Education." *Science* 281 (1998): 178–79.

Lederman, Leon. "Revolution in Science Education: Put Physics First!" *Physics Today* 54 (2001): 11–12.

National Research Council. *National Science Education Standards*. Washington, D.C.: National Academy Press, 1996.

Schmidt, William H., and Curtis C. McKnight. "What Can We Really Learn from TIMSS?" *Science* 282 (1998): 1830.

Schmidt, William H., Curtis C. McKnight, Leland S. Cogan, Pamela M. Jakwerth, and Richard T. Houang. *Facing the Consequences: Using TIMSS for a Closer Look at U.S. Mathematics and Science Education*. Boston: Kluwer Academic Publishers, 1999.

Schmidt, William H., Curtis C. McKnight, and Senta A. Raizen. *Splintered Vision: An Investigation of U.S. Science and Mathematics Education*. Boston: Kluwer Academic Publishers, 1997.

U.S. Department of Education, National Center for Education Statistics. *Pursuing Excellence: A Study of U.S. Eighth-Grade Mathematics and Science Teaching, Learning, Curriculum, and Achievement in International Context*. Washington, D.C.: U.S. Government Printing Office, 1996.

U.S. Department of Education, National Center for Education Statistics. *Pursuing Excellence: A Study of U.S. Fourth-Grade Mathematics and Science Achievement in International Context*. Washington, D.C.: U.S. Government Printing Office, 1997.

U.S. Department of Education, National Center for Education Statistics. *Pursuing Excellence: A Study of U.S. Twelfth-Grade Mathematics and Science Achievement in International Context*. Washington, D.C.: U.S. Government Printing Office, 1998.

ON THE ONENESS OF NATURE

Edward "Rocky" Kolb

There are deep and profound connections in nature. Connections between the largest things in the universe and the smallest, between the inner space of quantum physics and the outer space of the cosmos illustrate the oneness of nature.

THE ONENESS OF NATURE

In high schools, universities, and our learned academies, science is typically divided into departments such as physics, chemistry, biology, geology, astronomy, mathematics, and so forth. But these divisions are an artificial human construction; nature itself is not so cleanly divided. For example, understanding the largest things in nature requires understanding the smallest. The microworld and the macroworld are connected. One can't be understood without the other. Modern science is fulfilling the age-old search for oneness.[1]

Recent developments in the field of science called cosmology illustrate the connectivity of nature. Although there have been tremendous recent advances in cosmology, it is far from a recent enterprise. The American anthropologist George Murdock surveyed every culture and civilization ever known and concluded that they all share some common characteristics. One common

characteristic is that every culture had some sort of body adorn-
ment. But a more important characteristic is that every culture had
a shared view of the universe, that is, a cosmological model.

A cosmological model includes a story for the origin, size, and
composition of the universe. In primitive cultures, like ancient
European Paleolithic or modern American Fundamentalist, cosmology might be based on superstition, religion, philosophy,
myth, or some sort of revealed truth. Our cosmology in the
modern age, our modern view of the universe is based on science.
The fundamental goal of modern scientific cosmology is to base a
view of the universe on the foundation of the laws of nature discovered in terrestrial laboratories or extrapolated from experi-
mentally determined laws.

The modern big bang model is based on the principle that the
universe is basically the same everywhere, and appears the same
when viewed in any direction. This is known as the *Cosmological
Principle*. It is the ultimate extension of the Copernican revolution. In his revolutionary book of 1543, *De Revolutionibus*, Copernicus asserted that we do not occupy the center of the solar
system. In 1918, the American astronomer Harlow Shapley
demonstrated that our solar system is not at the center of our
galaxy. In 1924, Edwin Hubble discovered that our Milky Way is
but one galaxy out of billions in the observable universe. Finally,
the Cosmological Principle implies that our galaxy does not occupy
a privileged place among those billions of galaxies in the universe.

Cosmologists have often run afoul of authoritarian interests.
History has many examples, Galileo's troubles with the Inquisi-
tion being perhaps the most famous. A view of the universe is a
powerful thing. There were confrontations between cosmology
and authority even in the twentieth century. It is easy to see how
technology resulting from science impacts our lives, but profound
scientific ideas also have a tremendous impact on our culture and
society. A scientific principle like the Cosmological Principle may
sound harmless, but some view it as dangerous. The idea that the
same laws of nature apply everywhere in the universe was troubling to those in power in China during and after the time of the
Cultural Revolution. The fact that the same "rules" applied to

China as to the rest of the world was a subversive idea, and anyone who espoused the Cosmological Principle was persecuted. Dissident Chinese astrophysicist Fang Li-Zhi spoke of the Cosmological Principle in his acceptance speech for the Robert F. Kennedy Human Rights Award:[2]

> In the field of modern cosmology, the first principle is called the "Cosmological Principle." It states that the universe has no center; that it has the same properties throughout. Every place in the universe has, in this sense, equal rights. How can the human race, which has evolved in a universe of such fundamental equality, fail to strive for a society without violence and terror? How can we fail to build a world in which the rights of every human from birth are respected?

They are harmless-sounding words to some ears but subversive to others. Our cosmological model influences more than science. It is the canvas upon which we picture our place in the universe.

Our modern cosmological model incorporates the two profound discoveries of twentieth-century physics: general relativity and quantum mechanics. In the modern big bang model, the universe emerged from a state of high temperature and density some twelve to fifteen billion years ago. As the universe expanded and cooled, the rich and complex structure we see developed.

We understand the existence of galaxies and other large-scale structures as originating from the growth of small primordial seeds. If there were small initial seeds in an otherwise smooth distribution of matter, they would inexorably grow to become the galaxies and clusters we observe.

So the question of the origin of structure in the universe shifts to the question of the origin of the small primordial seeds. The remarkable recent discovery is that the seeds, and hence the origin of all structure in the universe, were planted by the action of microscopic quantum uncertainty in the early universe. This connection between the microphysical world of quantum mechanics and the macrophysical universe of galaxies is a wonderful example of the true unity of science.

The oneness of nature is not a new idea. About a century ago the great American naturalist John Muir eloquently wrote, "When one tugs at a single thing in nature, he finds it hitched to the rest of the universe." Galaxies, and even larger cosmological structures millions of light-years in size, are hitched to microscopic quantum uncertainty.

THE WINDS OF REVOLUTION

He spoke softly with a pipe clenched in his teeth. He mumbled a lot and English wasn't his native language. His closest colleagues had trouble understanding him speaking any language, even his native Danish. Some said that he had the rare linguistic ability to be unintelligible in four different languages.

It was hard to imagine a worse choice for a public speaker. But poor communication skills didn't cool the feverish excitement in the crowded lecture hall on the campus of the University of California at Berkeley. For in the audience that day were young students burning with the inner fire of committed revolutionaries, and they were listening to a decorated hero of the revolution.

In the tumultuous year of 1937, in the midst of a worldwide economic depression, with a civil war raging in Spain, fascists in power in Germany and Italy, and Japan and China about to go to war in Asia, the world seemed to be coming apart. Talks about revolutions were common on college campuses.

But the Danish speaker that day was an uncommon revolutionary. He was not concerned with the Russian Revolution that had occurred twenty years earlier, or any other political or social revolution. He spoke of a much more profound and far-reaching revolution. He spoke of a revolution that inevitably would shape society in ways a mere political revolution never could.

Niels Bohr was one of the great heroes of twentieth-century physics, decorated with the 1922 Nobel Prize for the "Study of Structure and Radiations of Atoms." He was at the forefront of the struggle to develop the laws of quantum mechanics, the seemingly bizarre set of rules that govern the behavior of matter on submicroscopic scales. Bohr was a leader of the revolution that over-threw our understanding of the laws of the natural world that had ruled physics since the time of Newton, and replaced them with a new order that included radical concepts such as the uncertainty principle, quantum levels, and quantum fluctuations.

When Bohr spoke in Berkeley that day in 1937, the revolution of quantum mechanics had not caused much of a ripple beyond a small community of physicists in the cloistered halls of universities and research labs. It would be decades before knowledge of the quantum world would change the everyday world by paving the way for the development of transistors, lasers, magnetic resonance imaging, digital computers, and all the products of our electronic society (and yes, nuclear weapons).

At the back of the crowded auditorium that day was a young physics graduate student named Philip Morrison. Now Institute Professor Emeritus at the Massachusetts Institute of Technology, a distinguished astrophysicist, and well-known science writer, he was then just another young face in the audience in awe of the great Bohr. The experience must have left quite an impression, because even sixty years after the event Morrison vividly recalled Bohr's lecture, as well as one particular question asked of Bohr after the lecture.

In the postlecture question period Bohr was asked what he thought of recent advances in cosmology. After all, the 1920s and 1930s were just as revolutionary for extragalactic astronomy as they were for physics. In 1924 Edwin Hubble extended the size of the known universe when he proved that *spiral nebulae* were distant galaxies like our own Milky Way. In 1929 Hubble's observations led to another fundamental discovery when he showed that distant galaxies are rushing away from us in an explosion of space we now call the big bang. By 1937 Hubble was charting the universe with the one hundred-inch Mt. Wilson Hooker Telescope and studying how galaxies are spread throughout space.

Lots of theoretical work in cosmology occurred in the 1920s and 1930s. By 1937 Aleksandr Friedmann in Russia and Georges Lemaitre in Belgium had proposed the big bang theory for the origin of the universe. Even Einstein, who had originally dismissed the possibility of an expanding universe, had by then accepted the idea of a big bang. It must have been puzzling that Bohr didn't mention cosmology in his lectures about the frontiers of physics.

Bohr's reply might have been even more surprising than his original omission of the subject. He answered that the astronomical observations and cosmological theories were important, but a true understanding of the origin and structure of the universe was impossible without first understanding how the laws of quantum mechanics operate on the fundamental particles. He said that until cosmology and particle physics could be brought together in the same context, he did not have much hope for real progress in cosmology.

Since quantum mechanics and particle physics are only important on atomic scales or smaller, it wasn't obvious that they could play a role in shaping something as large as a galaxy. A connection between the world of the ultralarge, the outer space of cosmology, and the world of the ultrasmall, the inner space of fundamental particles where the laws of quantum mechanics were important, must have seemed a pretty radical idea.

Although some of his fellow revolutionaries might have agreed with Bohr about the importance of quantum mechanics and particle physics in cosmology, I wonder if anyone there had an inkling of just how right he was. Once set in motion, revolutions take off in unanticipated directions. The pioneers of the quantum revolution couldn't tell where the revolution would lead. Bohr might have anticipated that in 1947, just ten years after his lecture, Ralph Alpher, George Gamow, and Robert Herman would study the nuclear physics of the early universe and predict the discovery of a background of microwave radiation. However, even Bohr might have had difficulty imagining that sixty years after his Berkeley lecture, cosmologists would believe that everything we see around us—galaxies, stars, planets, people, poodles, pigeons, and pond scum—started out as small quantum fluctuations in the primordial soup of the early universe.

The connections between quantum mechanics and galaxies started just two years after Bohr's lecture. In the mid-1930s Erwin Schrödinger, another founder of quantum mechanics, turned to cosmological issues. In the tumultuous period of 1938 to 1939, he was forced to leave his position in Graz, Austria, because of his left-wing politics and seek asylum in the Vatican. Those familiar with the life of Schrödinger understand why he was ill-suited for cloistered Vatican life. In 1939 he left the Vatican and accepted a temporary position in Ghent, Belgium. It didn't take the genius of Schrödinger to recognize that Belgium was not the ideal place to live with Germany and France heading toward war, so he accepted a position in Ireland at the Institute for Advanced Studies. In this disruptive and tragic period, amid all the turmoil, he wrote a remarkable paper, "Proper Vibrations of the Expanding Universe." In the introduction of the paper he wrote: ". . . proper vibrations [positive and negative frequency modes] cannot be rigorously separated in the expanding universe. . . . this is a phenomenon of outstanding importance. With particles it would mean production or annihilation of matter, merely by expansion, . . . Alarmed by these prospects, I have examined the matter in more detail."

In the conclusion, Schrödinger stated: "There will be a mutual adulteration of positive and negative frequency modes in the course of time, giving rise to . . . the 'alarming phenomenon.'"[3]

It is remarkable that in 1939 with the world around him engulfed in the flames of a world war, Schrödinger was "alarmed" by the creation of one particle in the universe every Hubble time (presently about 10,000 million years). By working with the world in turmoil, perhaps Schrödinger was following the precepts of an even earlier Graz cosmologist, Johannes Kepler. Because of political and religious instability during the period 1600–1630, Kepler had to flee Graz and find refuge in a series of temporary positions in Prague, Linz, Sagan, and Ratisbon. When reflecting on how he was tossed about on the seas of war and unrest, Kepler wrote, "When the storms rage around us and the state is threatened by shipwreck, we can do nothing nobler than to lower the anchor of our peaceful studies into the ground of eternity."[4]

In the face of the present world crisis, perhaps we as scientists can do nothing nobler than continue our peaceful studies and hope that one day they will be anchored into the ground of eternity.

What alarmed Schrödinger was the action of gravity on the vacuum of quantum uncertainty. Because of the uncertainty principle of Heisenberg, empty space is not a quiescent vacuum, but it foams and seethes with particles and antiparticles emerging from the vacuum existing for a brief period of time before disappearing back into the vacuum. These "virtual" particles are an intrinsic part of the vacuum. In quantum mechanics, nothing is something.

In 1939 Schrödinger realized that the strong gravitational field of the expanding universe could convert virtual vacuum particles into real particles. This is a similar phenomenon to Hawking radiation in the vicinity of a black hole. In 1974 Stephen Hawking realized that black holes aren't so black after all. In the vicinity of the black hole horizon, one of the virtual particles can fall into the black hole and its partner escape to infinity. In this way, particles may be extracted from the vacuum. In the language of Schrödinger, the positive and negative frequency modes are "adulterated."

In the rapid expansion of the early universe, particles and antiparticles can appear from the vacuum (nothing) and be pulled apart by the expansion of space before they have a chance to annihilate. This may appear to be much ado about nothing, but it implies that the expansion of the universe will create particles and destroy perfect homogeneity and create the primordial seeds that will grow to become all we see in the universe.

If this idea is correct, then all structure we see in the universe today (galaxies, galaxy clusters, as well as the pattern of temperature fluctuations) are actually patterns of quantum fluctuations.

When we first learned of quantum mechanics in the sixth grade, we were told that it is impossible to see the effects of quantum mechanics because they are too small. However, the rapid expansion of the universe during inflation stretched microphysical fluctuations to scales as large as the entire observable universe. The fact that quantum fluctuations can be seen using telescopes (in addition to the most powerful microscopes, particle accelerators) may be the most elemental example of the inner space/outer space connection.

IMPORTANT EVENTS IN COSMOLOGY, FROM THE BIG BANG TO THE BIRTH OF LEON LEDERMAN

Understanding the inner space/outer space connection helps complete the picture for the complete history of the universe. If different epochs in the history of the universe can be thought of as different movements in the cosmic symphony, the score is given in Table 1.

Table 1. The Cosmic Symphony (Harmonice Mundi)

Tempo	Movement	Epoch	Relic
Pizzicato	String	10^{-43} seconds?	????
Prestissimo	Inflation	10^{-35} seconds?	Seeds of structure
Presto	Radiation	The first 10,000 years	Abundances of the light elements
Allegro	Matter	10,000 years after the bang	Distant quasars and galaxies
Andante	Vacuum (inflation)	A billion years ago	Present acceleration of the universe
Largo	Antiquity	15 July 1922	Leon Lederman

My string-theorist friends tell me (or, if I had any friends who were string theorists, they would tell me) that the string section may dominate the first movement of the Cosmic Symphony if on the smallest scales there is a fundamental stringiness to elementary particles. If this were true, then the first movement in the Cosmic Symphony would have been a pizzicato movement of vibrating strings about 10^{-43} seconds after the bang. There is basically nothing known about the stringy phase, if indeed there was one. We do not yet know enough about this era to predict any surviving relics.

The earliest movement for which we can detect an echo is the inflationary phase. The inflationary movement probably followed the string movement and lasted approximately 10^{-35} seconds. The best information we have of the inflationary phase is from the quantum seeds planted during that epoch, which can be seen today in the form of galaxies.

We do know that the universe was radiation-dominated for almost all of the first 10,000 years. The best-preserved relics of the radiation-dominated era are the light elements. The light elements were produced in the radiation-dominated universe one second to three minutes after the bang.

Very distant quasars and galaxies give us a picture of the early matter-dominated era. Structure developed from small primordial seeds during that time.

Finally, if recent astronomical observations are correctly interpreted, the expansion of the universe is accelerating today. This would mean that the universe has recently embarked on another inflationary era, but with the Hubble expansion rate much less than the rate during the first inflationary era.

A final significant cosmological event occurred on 15 July 1922. The fossil of this event is Leon Lederman, who is an inspiration to cosmologists throughout the civilized world and many parts of Warrenville, Illinois. In 1983 Leon Lederman planted cosmological seeds of his own at Fermi National Accelerator Laboratory. Leon's own revolutionary book, *Dei Particilus*, along with Copernicus's *De Revolutionibus*, Galileo's *Dialogo Sopra i Due Massimi Sistimi del Mondo*, and Newton's *Principia*, is a landmark contribution to the inner space/outer space synthesis.[5]

NOTES

1. How did the New York hot dog vendor help the Buddhist monk attain enlightenment? He made him one with everything.

2. A friend delivered the speech in Washington because Fang was in exile in the American Embassy in Beijing at the time.

3. Erwin Schrödinger, "The Proper Vibrations of the Expanding Universe," *Physica* 6 (1939): 899.

4. Johannes Kepler, in a letter to Bartsch, as found in *Gesammelte Werke* (C. H. Beck'sche Verlagsbuchhandlung, München).

5. I am grateful for the support of the United States Department of Energy and NASA (NAG5-10842). I am also grateful to Leon Lederman for inspiration and providing the opportunity to pursue the study of the universe.

SCIENTIFIC INQUIRY AND NATURE OF SCIENCE AS A MEANINGFUL CONTEXT FOR LEARNING IN SCIENCE

Norman G. Lederman

INTRODUCTION

Science educators have historically been concerned with students' ability to apply scientific knowledge to make informed decisions regarding personal and societal issues. The ability to use scientific knowledge to make informed personal and societal decisions is the essence of what contemporary science educators and reform documents define as scientific literacy. The most recent reform visions of note have been the *National Science Education Standards* and *Project 2061* (AAAS) and these efforts stress the importance of conceptual understanding of the overarching ideas in science (e.g., cause and effect, equilibrium, structure and function, cycles, scale). The phrase "less is more" has often been invoked to communicate the desire that instructional time focus on in-depth understanding of a reduced set of unifying scientific concepts.

Current reform documents also stress an increased emphasis in two areas that make them significantly different from previous efforts: nature of science and scientific inquiry. Helping students develop adequate conceptions of nature of science (NOS) and scientific inquiry has been a perennial objective in science education (AAAS 1990,

1993; Klopfer 1969; NRC 1996; NSTA 1982) extending back to the beginning of the twentieth century (CASMT 1907). "The longevity of this educational objective has been surpassed only by the longevity of students' inability to articulate the meaning of the phrase *nature of science*, and to delineate the associated characteristics of science" (Lederman and Niess 1997) or *scientific inquiry*.

Consequently, it is only natural to ask whether there are reasons to believe that the recent reforms in science education are more likely to impact students' understandings than their predecessors. Two critical and interrelated omissions that have typified previous efforts are, unfortunately, evident in the more recent reform documents. There is not, and there has not been, a concerted professional development effort to clearly communicate first, what is meant by *NOS* and *scientific inquiry* and second, how a functional understanding of these valued aspects of science can be communicated to K–12 students. Perhaps the lack of professional development related to NOS and scientific inquiry is a consequence of the misunderstanding that NOS and scientific inquiry are cognitive outcomes of less importance than "traditional" subject matter. In reality, however, it is NOS and scientific inquiry that provide the context for the subject matter specified in the *Standards* and other reform documents. In the following sections, I will clarify the meaning of *NOS* and *scientific inquiry*. I will also delineate several misconceptions promoted (or ignored) by reform efforts. It will further be argued that without explicit/reflective instructional attention to NOS and scientific inquiry, students will continue to learn science subject matter in a context-free environment. Such an environment does not permit the in-depth conceptual understanding of science subject matter advocated in the various visions of reform and will not help create a populace that can be considered scientifically literate. A functional understanding of NOS and scientific inquiry by teachers is clearly prerequisite to any hopes of achieving the vision of science teaching and learning specified in the various reform efforts.

WHAT IS NATURE OF SCIENCE?

The phrase *nature of science* typically refers to the epistemology of science, science as a way of knowing, or the values and beliefs inherent to scientific knowledge and its development (Lederman 1992). There is a recognized lack of consensus about specific aspects of NOS among philosophers of science, historians of science, scientists, and science educators. This lack of consensus, however, should be neither disconcerting nor surprising given the multifaceted nature and complexity of the scientific endeavor. It is my view, however, that many of the disagreements that continue to exist are irrelevant to K–12 instruction. The issue of the existence of an objective reality as compared to phenomenal realities is a case in point. I argue that there is an acceptable level of generality regarding NOS that is accessible to K–12 students and relevant to their daily lives. Moreover, at this level, little disagreement exists among philosophers, historians, and science educators. Among the characteristics of the scientific enterprise corresponding to this level of generality are that scientific knowledge is tentative (subject to change); empirically based (based on and/or derived from observations of the natural world); subjective (theory-laden); necessarily involves human inference, imagination, and creativity (involves the invention of explanations); and is socially and culturally embedded.

One additional aspect of NOS, closely related to the distinction between observation and inference, is the distinction between scientific theories and laws. Laws are *statements or descriptions of the relationships* among observable phenomena. Boyle's law, which relates the pressure of a gas to its volume at a constant temperature, is a case in point. Theories, by contrast, *are inferred explanations* for observable phenomena. The kinetic molecular theory, which explains Boyle's law, is one example.

Professional development efforts designed for teachers must not conclude, as they have in the past, with the development of adequate teacher understandings. The research is quite clear that teachers' understandings do not automatically translate into classroom practice. Certainly, teachers must have an in-depth understanding of what they are expected to teach. However, professional development efforts must also emphasize how teachers can successfully facilitate the development of students' understandings of NOS.

WHAT IS SCIENTIFIC INQUIRY?

Although closely related to science processes, scientific inquiry extends beyond the mere development of process skills such as observing, inferring, classifying, predicting, measuring, questioning, interpreting, and analyzing data. Scientific inquiry includes the traditional science processes but also refers to the combining of these processes with scientific knowledge, scientific reasoning, and critical thinking to develop scientific knowledge. From the perspective of the *National Science Education Standards* (NRC 1996), students are expected to be able to develop scientific questions and then design and conduct investigations that will yield the data necessary for arriving at conclusions for the stated questions, as well as develop understandings of the process and its implications for the knowledge claims developed. The *Benchmarks for Science Literacy* (AAAS 1993) are a bit less ambitious as they do not advocate that all students be able to design and conduct investigations in total. Rather, it is expected that all students at least be able to understand the logic of an investigation and be able to critically analyze the claims made from the data collected. Scientific inquiry, in short, refers to the systematic approaches used by scientists in an effort to answer their questions of interest.

Precollege students, and the general public for that matter, believe in a distorted view of scientific inquiry that has resulted from schooling, the media, and the format of most scientific reports. This distorted view is called *The Scientific Method*, that is, a fixed set and sequence of steps that all scientists follow when attempting to answer scientific questions. The visions of reform, however, are quick to point out that there is no single fixed set or sequence of steps that all scientific investigations follow. The contemporary view of scientific inquiry advocated is that the questions guide the approach and the approaches vary widely within and across scientific disciplines and fields.

The perception that a single scientific method exists owes much to the status of classical experimental design. Experimental designs very often conform to what is presented as "The Scientific Method," and the examples of scientific investigations presented in science textbooks most often are experimental in nature. The problem is that experimental research is not representative of scientific investigations as a whole.

Scientific inquiry has always been ambiguous in its presentation within science education reforms. In particular, inquiry is perceived in three different ways in the current reforms. It can be viewed as a set of skills to be learned by students and combined in the performance of a scientific investigation. It can also be viewed as a cognitive outcome that students are to achieve. In particular, the current visions of reform are very clear (at least in written words) in distinguishing between the performance of inquiry (for example, what students will be able to do) and what students know about inquiry (for example, what students should know). For example, it is one thing to have students set up a control group for an experiment; it is another to expect students to understand the logical necessity for a control within an experimental design. The third use of *inquiry* in reform documents relates strictly to pedagogy and further muddies the water. In particular, current wisdom advocates that students best learn science through an inquiry-oriented teaching approach. It is believed that students will best learn scientific concepts by doing science. In this sense, scientific inquiry is viewed as a teaching approach used to communicate scientific knowledge to students as opposed to an educational outcome that students are expected to learn about and learn how to do.

COMMUNICATING FUNCTIONAL UNDERSTANDINGS OF THE NOS

The tone of my discussion implies that science education reforms, currently and in the past, have mishandled NOS and scientific inquiry. It has been assumed that teachers understand both of these important aspects of science and little professional development has been planned or provided. There is an additional critical flaw in the various reforms' approach to the teaching of nature of science and scientific inquiry. It is this critical flaw that has existed since the beginning of the science education community's recognition of the importance of scientific inquiry and NOS as important educational outcomes.

Two general approaches have been advocated by reform documents and the science education literature to enhance students' and teachers' understandings of NOS and/or scientific inquiry. The first approach, labeled here as an *implicit* approach, suggests that by "doing science" students will also come to understand NOS and scientific inquiry (Lawson 1982; Rowe 1974). This approach was adopted by most of the curricula of the 1960s and 1970s that emphasized hands-on, inquiry-based activities and/or process-skills instruction. Research studies have indicated that the implicit approach was not effective in enhancing students' and teachers' understandings of NOS or scientific inquiry (for example, Durkee 1974; Haukoos and Penick 1985; Riley 1979; Spears and Zollman 1977; Trent 1965; Troxel 1968).

The second approach, the *historical* approach (one that is strongly recommended by the *National Science Education Standards*), suggests that incorporating the history of science (HOS) in science teaching can serve to enhance students' views of NOS. However, a review of the efforts that aimed to assess the influence of incorporating HOS in science teaching (Klopfer and Cooley 1963; Solomon et al. 1992; Welch and Walberg 1972; Yager and Wick 1966) indicates that evidence concerning the effectiveness of the historical approach is, at best, inconclusive. And, most recently, the work of Abd-El-Khalick (1998) has indicated that specific courses in the history and/or philosophy of science have little impact on students' understanding of NOS and scientific inquiry.

An alternative approach to the two often noted in the reforms suggests that the goal of improving students' views of the scientific endeavor "should be planned for instead of being anticipated as a side effect or secondary product" of varying approaches to science teaching (Akindehin 1988). This *explicit/reflective* approach uses instruction geared toward various aspects of NOS or scientific inquiry and utilizes elements from the history and philosophy of science to improve learners' views of NOS. In general, relative to the implicit and historical approaches, the explicit/reflective approach has been more effective in helping learners achieve enhanced understandings of NOS and scientific inquiry (for example, Akindehin 1988; Billeh and Hasan 1975; Carey and Strauss 1968, 1970; Jones 1969; Lavach 1969; Ogunniyi 1983; Olstad 1969).

A functional understanding of NOS and/or scientific inquiry is best facilitated through an explicit/reflective approach. I cannot overemphasize the importance of taking time, at the conclusion of any activity, to *explicitly* point out to students (or promote student discussion about) the aspects of NOS and scientific inquiry that are highlighted. To encourage reflection, teachers must discuss with students the implications such aspects of NOS and scientific inquiry have for the way they view scientists, scientific knowledge, and the practice of science.

CONCLUDING REMARKS

I began this discussion by distinguishing current reform efforts in science education from their predecessors with respect to the heightened interest and emphasis on scientific inquiry and NOS. The primary reason for this heightened, but certainly not new, advocacy is the belief that students need to develop in-depth understandings of how scientific knowledge is generated and the implications this has for the status of the knowledge. Science educators have come to believe that if students understand the source and limits of scientific knowledge they will be better equipped to make informed decisions about personal and societal issues that are scientifically based. In short, understandings of NOS and scientific inquiry are believed to be critical and essential components of the modern-day battle cry of "scientific literacy."

With respect to students' achieving in-depth understanding of subject matter, it can be argued that such a goal is unachievable unless students understand NOS and scientific inquiry. For example, can it be said that a student truly understands the concept of a gene if she does not realize that a gene is a construct invented to explain experimental results? Does the student who views genes as possessing physical existence analogous to pearls on a necklace possess an in-depth understanding of the concept? Does the student who is unaware that the atom (as pictured in books) is a scientific model used to explain the behavior of matter and that it has not been directly observed have an in-depth understanding of the atom?

Misconceptions about the scientific validity of biological evolution commonly appear in the media and courts of law. Many of these misconceptions relate to whether evolution is a testable scientific theory. The arguments against the validity of evolution usually proceed to point out that evolution cannot be tested using the scientific method. Therefore, evolution cannot be a valid scientific theory. Many feel that the problem is at least partially created by the public's misunderstanding of scientific inquiry and/or scientific theory. These few examples should make it clear that understanding of NOS and scientific inquiry provide a guiding framework and context for scientific knowledge. Without an understanding of how scientific knowledge is derived and the implications the process of derivation has for the status and limitations of the knowledge, all students can ever hope to achieve is knowledge without context. Context is necessary for students to understand what the knowledge means. In short, lack of context is the equivalent to playing a game of chess without knowing the rules of the game. Unless students can derive meaning for the scientific knowledge they acquire, there is little hope that they can use their knowledge to make informed decisions.

Over the years, we have come to realize that students cannot meaningfully learn a long laundry list of terms, vocabulary, and factoids. We have also recognized the sensibility of attempting to focus our educational efforts on fewer, unifying themes/concepts. However, we continue to fail at providing students with the most important organizing themes of all, NOS and scientific inquiry. Despite volumes of research we continue to believe that students will come to understand scientific inquiry and NOS simply by "doing science." Such an expectation is equivalent to assuming individuals will come to understand the mechanism of breathing simply by breathing. Obviously, this is not the case. Doing science is certainly a start, but students need to reflect on what it is they are doing. They need to be engaged in discussions of why scientific investigations are designed in certain ways. Students need to discuss the assumptions inherent to any scientific investigation and the implications these assumptions have for the results. Furthermore, students need to discuss that science is done by humans and to discuss the implications this has for the knowledge that is produced. NOS and scientific inquiry need to be addressed explicitly during science instruction. They need to be given status equal to that of traditional subject matter. Without such explicit/reflective instructional attention, students will continue to learn subject matter without context and the visions of reform in science education will progress no further than they have in the past.

REFERENCES

Abd-El-Khalick, F. *The Influence of History of Science Courses on Students' Conceptions of the Nature of Science.* Unpublished doctoral diss., Oregon State University, 1998.

Akindehin, F. "Effect of an Instructional Package on Preservice Science Teachers' Understanding of the Nature of Science and Acquisition of Science-Related Attitudes." *Science Education* 72 (1988): 73–82.

American Association for the Advancement of Science (AAAS). *Science for All Americans.* New York: Oxford University Press, 1990.

American Association for the Advancement of Science (AAAS). *Bench-*

marks for Science Literacy: A Project 2061 Report. New York: Oxford University Press, 1993.

Billeh, V. Y., and O. E. Hasan. "Factors Influencing Teachers' Gain in Understanding the Nature of Science." *Journal of Research in Science Teaching* 12 (1975): 209–19.

Carey, R. L., and N. G. Stauss. "An Analysis of the Understanding of the Nature of Science by Prospective Secondary Science Teachers." *Science Education* 52 (1968): 358–63.

———. "An Analysis of Experienced Science Teachers' Understanding of the Nature of Science." *School Science and Mathematics* 70 (1970): 366–76.

Central Association of Science and Mathematics Teachers (CASMT). "A Consideration of the Principles That Should Determine the Courses in Biology in the Secondary Schools." *School Science and Mathematics* 7 (1907): 241–47.

Durkee, P. "An Analysis of the Appropriateness and Utilization of TOUS with Special Reference to High-Ability Students Studying Physics." *Science Education* 58 (1974): 343–56.

Haukoos, G. D., and J. E. Penick. "The Effects of Classroom Climate on College Science Students: A Replication Study." *Journal of Research in Science Teaching* 22 (1985): 163–68.

Jones, K. M. "The Attainment of Understandings about the Scientific Enterprise, Scientists, and the Aims and Methods of Science by Students in a College Physical Science Course." *Journal of Research in Science Teaching* 6 (1969): 47–49.

Klopfer, L. E. "The Teaching of Science and the History of Science." *Journal of Research for Science Teaching* 6 (1969): 87–95.

Klopfer, L. E., and W. W. Cooley. "The History of Science Cases for High Schools in the Development of Student Understanding of Science and Scientists." *Journal of Research for Science Teaching* 1 (1963): 33–47.

Lavach, J. F. "Organization and Evaluation of an Inservice Program in the History of Science." *Journal of Research in Science Teaching* 6 (1969): 166–70.

Lawson, A. E. "The Nature of Advanced Reasoning and Science Instruction." *Journal of Research in Science Teaching* 19 (1982): 743–60.

Lederman, N. G. "Students' and Teachers' Conceptions of the Nature of Science: A Review of the Research." *Journal of Research in Science Teaching* 29 (1992): 331–59.

Lederman, N. G., and M. Niess. "The Nature of Science: Naturally?" *School Science and Mathematics* 97 (1997): 1–2.

National Research Council (NRC). *National Science Education Standards.* Washington, D.C.: National Academic Press, 1996.

National Science Teachers Association (NSTA). *Science-Technology-Society: Science Education for the 1980s* (an NSTA position statement). Washington, D.C.: National Science Teachers Association, 1982.

Ogunniyi, M. B. "Relative Effects of a History/Philosophy of Science Course on Student Teachers' Performance on Two Models of Science." *Research in Science and Technological Education* 1 (1983): 193–99.

Olstad, R. G. "The Effect of Science Teaching Methods on the Understanding of Science." *Science Education* 53 (1969): 9–11.

Riley, J. P., II. "The Influence of Hands-On Science Process Training on Preservice Teachers' Acquisition of Process Skills and Attitude Toward Science and Science Teaching." *Journal of Research in Science Teaching* 16 (1979): 373–84.

Rowe, M. B. "A Humanistic Intent: The Program of Preservice Elementary Education at the University of Florida." *Science Education* 58 (1974): 369–76.

Solomon, J., J. Duveen, L. Scot, and S. McCarthy. "Teaching about the Nature of Science through History: Action Research in the Classroom." *Journal of Research in Science Teaching* 29 (1992): 409–21.

Spears, J., and D. Zollman. "The Influence of Structured versus Unstructured Laboratory on Students' Understanding of the Process of Science." *Journal of Research in Science Teaching* 14 (1977): 33–38.

Trent, J. "The Attainment of the Concept 'Understanding Science' Using Contrasting Physics Courses." *Journal of Research in Science Teaching* 3 (1965): 224–29.

Troxel, V. A. *Analysis of Instructional Outcomes of Students Involved with Three Sources in High School Chemistry.* Washington, D.C.: U.S. Department of Health, Education, and Welfare, Office of Education, 1968.

Welch, W. W., and H. J. Walberg. "A National Experiment in Curriculum Evaluation." *American Educational Research Journal* 9 (1972): 373–83.

Yager, R. E., and J. W. Wick. "Three Emphases in Teaching Biology: A Statistical Comparison of Results." *Journal of Research in Science Teaching* 4 (1966): 16–20.

In Praise of Audacity

Tackling the Big Problems

Shirley M. Malcom

I f any prologue is fitting for an essay about Dr. Leon Lederman, it is this one drawn from Dr. Benjamin E. Mays (1895–1984). Dr. Mays was president of Morehouse College from 1940 to 1967. Dr. Martin Luther King Jr. called Dr. Mays his "spiritual and intellectual mentor." Dr. Mays inspired dreams and dreamers.

The tragedy of life doesn't lie in not reaching your goal. The tragedy lies in having no goal to reach. It isn't a calamity to die with dreams unfilled, but it is a calamity not to dream. It is not a disgrace not to reach the stars, but it is a disgrace to have no stars to reach for. Not failure, but low aim, is a sin.

This essay in honor of Leon Lederman recognizes the work that he has done to improve the quality of K–12 science and mathematics education in the United States and, indeed, throughout the world. While this passion is a major aspect of his current professional interest, even while director of Fermi National Accelerator Laboratory (Fermilab) he developed and supported education programs to support student interest in physics. It was largely through his encouragement and advocacy that the Illinois Mathematics and Science Academy (IMSA) was established by the governor and state legislature to support the nurturing of talented students for that state. But the audaciousness appears in his willingness to rise to the challenge, to be bold in his vision, and to defy traditions, if necessary, to share science broadly.

RISING TO THE CHALLENGE

In a 1987 speech, then U.S. Secretary of Education William Bennett characterized Chicago as having the worst school district in the country. While others might have just disparaged the remark, Lederman took the challenge personally. Unwilling to write off over 400,000 students, the majority of which were African American, Latino, and poor, Lederman garnered support to establish the Teachers Academy for Mathematics and Science (TAMS). Building on the research that connected student performance to teacher knowledge and skills, TAMS's strategy and principles were straightforward:

- Teachers cannot teach what they do not know how to teach. Meet their needs; engage with mathematics and science as subjects they have never been taught. Use inquiry strategies known to be effective with children. Teachers need a safe environment to acquire critical knowledge and skills.
- Professional development must embrace the entire school, with buy-in from principals and staff as well as teachers.
- Focus on elementary schools to ensure that children are not turned off early to mathematics and science learning.
- Focus on schools that are ready for change and that have the type of demographics that characterize the system. Evidence must be presented to reinforce the notion that all children are capable of learning to very high levels.
- Evaluate the programs; collect data to inform the interventions and to document efforts.
- Make interventions research-based. Consult widely, *drawing on the wisdom of practice.*

The boldness of Lederman's initiative lies in the willingness to undertake these efforts in a large urban system, to begin with student populations that have historically underperformed in science and mathematics—where expectations for their performance may be low—with teachers often underprepared in science and mathematics, and to see this as an opportunity.

His vision was eventually to take on the entire system, but there was never enough funding in place to make that leap. Some might characterize these early efforts as overly ambitious, but Lederman's instincts to do this work at a scale commensurate with the size of the challenges were on target. Lederman used his celebrity as a Nobel laureate and eminent scientist to gain access to people who might support these education initiatives. He valued the need to get started, using the best information available, and then fine-tuning the intervention once more was known. Business management programs might refer to this strategy as "ready, fire, aim." Those of us who have experienced the Lederman tsunami simply refer to it as "typical Leon."

Recent research on system-wide approaches to science and mathematics education reform and on successful strategies for professional education of teachers bears out the wisdom in the design that he favored. In K–12 education, systems have a way of absorbing and consuming efforts at incremental change. Only through boldness and a commitment to taking on the entire structure can one hope to succeed.

SCIENCE FOR ALL

From its inception, IMSA has been committed to a diverse student population and has been aggressive in increasing the numbers of historically underrepresented populations in mathematics and science. Rather than declaring minority students to be incapable of performing in—or uninterested in taking advantage of—the enriched programs of IMSA, and rather than lowering the bar for admission, the leadership of the school reached out to address other barriers that such students often face:

- Lack of information about program opportunities,
- Need for more emphasis on outreach and early identification,
- The need for better early preparation.

The fundamental values of the institution reflect a belief in the need to nurture talent wherever it might be located. These values apply broadly to Lederman's core beliefs, whether to provide opportunities for children born of poverty or to provide opportunities for research to scientists from developing countries. To live these values, to draw others who share these values into one's sphere—this is the legacy of Leon Lederman. This legacy of diversity is reflected in the faces of IMSA, increasingly in the face of Fermilab, and most certainly in the faces of the teachers and students who have benefited from the TAMS experience.

CHALLENGING TRADITION

Physics remains in many ways the science most inaccessible to the general public. Only a small fraction of high school graduates ever take a formal course in physics, largely due to its placement in the U.S. high school curriculum as the culminating course in most science programs. According to the National Center for Education Statistics (Legum 2001), less than 25 percent of all 1994 high school graduates completed a course in physics (see Table 1). This compares to 59 percent for chemistry and 93 percent for biology. In 1998 almost 29 percent of all high school graduates took a course in physics. When these course-taking figures are disaggregated by race and sex, the lesser likelihood of female and underrepresented minority students taking physics emerges.

Because of the way that physics is traditionally taught, according to results from the Third Mathematics and Science Study (TIMSS), even U.S. students who take high school physics classes at the highest level appear to leave school with a lower level of understanding of fundamental concepts in physics compared with students from other countries (U.S. Department of Education 1998).

Many school systems have few teachers certified to teach

Table 1. Course-Taking Patterns in Science for 1994 High School Graduates

	Biology	Chemistry	Physics
All	92.7	60.4	28.8
Male	91.4	57.1	31.7
Female	94.1	63.5	26.2
White	93.7	63.2	30.7
Asian/Pacific	92.9	72.4	46.9
Black	92.8	54.3	21.4
Hispanic	86.5	46.1	18.9
Native American or Alaskan Native	91.3	46.9	16.2

(Legum 2001)

physics. This is especially true in large urban districts. There is a low and declining number of physics majors in colleges and universities. Recent data from the American Institute of Physics indicates that the number of individuals receiving physics bachelor's degrees in 1999 was at a forty-year low, despite an increase in the total numbers of bachelor's degrees awarded. And, perhaps not surprisingly, the participation of women and minorities has been more depressed in physics than in other subject areas requiring similar quantitative preparation, areas such as engineering (Mulvey 2001).

Even though more than 90 percent of all high school graduates complete a course called biology, it is unlikely that the important emerging ideas and concepts of this field are understandable to students since they rely on knowledge usually gained in courses called chemistry and physics. Lederman, recognizing the apparent incongruity in this, initiated a movement called "physics first" (Lederman 1998). His proposition was to alter the order in which the sciences are introduced in high school, beginning with physics as a more phenomenon-based field. While those proposing a focus on standards argued that this would be unlikely to solve the problem of science course-taking in high school, many people in physics rejected the notion of being moved from the capstone position. If physics came first, they argued, it would need to be taught to all, including those believed by many to be incapable of learning it.

The pragmatists pointed to the problems of mobilizing an army of teachers to provide such a course, of developing and deploying courses at the appropriate level, of ensuring that mathematics, at least at the level of algebra, was completed by the time students would be taking the physics, and so on. Others, including advocates of science for all, saw an opportunity to support the idea of science and mathematics learning to high levels for minority and female students. Recognizing that somewhere someone must be doing it this way, Lederman went in search of the mavericks. And where he found them he found greater numbers of all students taking much more science courses of all kinds. Encouraged by these findings he pressed the issue.

Lederman received a small grant from the U.S. Department of Education to bring together a small group of practitioners, policy people, believers, and agnostics to imagine what science might look like over the entirety of a high school program. He modified his original notion to propose teaching of Science I, II, and III, that is, courses that would be, in order, mostly physics, mostly chemistry, and then mostly biology. Such a scheme would leave room for integrating concepts from earth and space science, technology and the like, as well as allowing students to pursue any of these fields in more depth. The notion was not necessarily to create more physicists (though that might be one outcome) but to make everyone much more science savvy. He would be the first person to acknowledge that this scheme was not *the solution* but that it represented a response to a current menu of high school science that poorly serves most students, where most students learn no physics and learn little about the modern biology that will likely inform their decision-making for a lifetime.

Lederman stirred up the dialogue, adding a dimension for those of us in search of transitional strategies to help bridge between what we have and what we believe that we really need.

And folks are still talking. Only boldness can push us away from incrementalism. And only the ultimate insider (a physicist, no less) can challenge a tradition of physics as an elite course, a proposal that might be considered to border on heresy.

TAKING ON THE WORLD

A lack of quality science and mathematics education is not a problem only for urban school systems in the United States. And supporting teachers to incorporate hands-on, inquiry-based approaches to teaching and learning science may just not be an appropriate response in the United States.

French Nobel laureate Georges Charpak visited TAMS in Chicago as well as a number of schools associated with the academy. He immediately recognized this model as one that could be adapted to address similar concerns about science learning in France. Charpak took video footage back to France and arranged for appearances in France with Lederman to begin a campaign for using inquiry-based, hands-on approaches to teaching and learning science. Charpak was able to engage and recruit members of the French Academy of Sciences to undertake this effort. What began as a pilot program is now underway throughout the country. La Main à la Pâte, a program supporting hands-on, inquiry-based science in primary schools in France (http://www.inrp.fr/lamap/reseau/interna/map_en.htm) is revolutionizing not only teaching and learning science in these schools but also is building a working relationship between the education and scientific communities.

La Main à la Pâte has been shared with countries all over the world. The major working documents were translated into Chinese, and a collaboration was established to assist in presenting the model in China.

Lederman has been able to employ "viral marketing" for use of inquiry-based, hands-on science for primary-level students. Along with colleagues from France, he has supported this as a world effort that has reached countries in Africa, Asia, and the

Americas. Perhaps centralization of school curricula in other countries of the globe allows more rapid adoption of core educational strategies, but everywhere, as in Chicago, the hard work of implementation almost always follows a slower timeline. Persistence must be a component of every scheme, along with the development of resources to support the work in schools.

THE BOTTOM LINE

The theme of this essay—taking on the problems worthy of one's time, effort, and energy—was an easy one to select. And the examples were easy to find. Lederman gives assignments to all of his friends and colleagues. And, generally, these are taken on because of our belief that he has already done much of the vetting as to their significance and importance.

When we (hopefully) gather for the Lederman centennial year there will still be a long "to do" list for K–12 science and mathematics education. Some of the unfinished business will likely be the same (that is, implementation). Hopefully, there will be more hands set to the task of reducing the list; hopefully, these will come from the many different sectors who must own the problems. If that is to happen, much of it will be due to the example set by Leon Lederman, who never shies away from hard work and big problems.

REFERENCES

Lederman, L. M., M. Bardeen, W. Freeman, S. Marshall, B. Thompson, and M. J. Young. *ARISE: American Renaissance in Science Education Three-Year High School Science Core Curriculum: A Framework*. Batavia, Ill.: Fermilab-TM-2051, 1998.

Legum, S., U.S. Dept. of Education, Office of Educational Research and Improvement, et al. *The 1998 High School Transcript Study Tabulations: Comparative Data on Credits Earned and Demographics for 1998, 1994, 1990, 1987, and 1982 High School Graduates*.

Washington, D.C.: U.S. Dept. of Education, Office of Educational Research and Improvement, 2001.

Mulvey, Patrick J., and Starr Nicholson. "Enrollments and Degrees Report." *Institute of Physics* (2001) AIP report R-151.37.

U.S. Department of Education, National Center for Education Statistics. *Pursuing Excellence: A Study of U.S. Twelfth-Grade Mathematics and Science Achievement in International Context*. Washington, D.C.: U.S. Government Printing Office, 1998.

THE MAGIC OF SCIENCE

Michael S. Turner

In January 1998 I came to Washington, D.C., under the saddest of circumstances. My mentor and close friend David N. Schramm had died tragically a few weeks earlier, and I was there to lead a memorial session at the American Astronomical Society. After the session, I was wandering aimlessly around a room filled with poster papers. Saul Perlmutter, the leader of the Supernova Cosmology Project at Berkeley, grabbed me and asked if I wanted to see something interesting.

His presentation had been relegated to a poster paper because it arrived too late to be included in the regular program. What he showed me changed the direction of my research and recharged my science batteries. It also changed the course of astronomy and physics. This is the special magic of science—Nature surprising us.

For more than a decade, Perlmutter and his team had been trying to measure the slight slowing of the expansion of the universe due to the cosmic gravitational pull of all the stuff within it. Since Edwin Hubble's discovery of the expansion itself, measuring this slowing has been a holy grail of cosmology. Its detection would confirm our understanding of how gravity operates on the grandest scales, while at the same time allowing us to determine the fate of the universe. Little or no slowing would mean expansion forever; lots of slowing would mean the eventual recollapse of the universe to a big crunch.

The trick in all of this is to use very distant objects to peer back in time. Because the time it takes light to travel to us from distant objects is significant, such objects are seen at an earlier epoch. The catch is finding objects suitable to be cosmic mileposts to chart the expansion history. For this purpose astronomers use so-called standard candles—objects of known intrinsic brightness—that can be seen across the universe. The brightest of these, Type Ia supernovae (white dwarf stars weighing about 40 percent more than the sun that undergo complete thermonuclear explosion) outshine the galaxies that host them for the month or so of their glory. By carefully measuring how bright distant supernovae appear on Earth, their distances can be inferred (the more distant they are, they fainter they appear). Using this technique, a decade ago, Perlmutter and his team set out to trace the expansion history of the universe back ten billion years.

All of this is easier said than done. In a typical galaxy, supernovae occur only once every couple hundred years. Thousands of galaxies have to be monitored to harvest a handful of supernovae each year. Using the biggest astronomical digital camera, built by Tony Tyson of Bell Labs, and software they developed to compare images of the sky taken at two different times, Perlmutter and his team searched for the telltale signature of a supernovae: a difference in the light profile of a galaxy between the two images. After many ups and more downs, including a chorus of "you'll never succeed" sung by many astronomers, in 1998 they announced their amazing result: The universe is speeding up, not slowing down. The discovery was a stunning reversal. Physicists and astronomers alike wanted to know how this could be. Not wanting to be left out of the fun, I turned the focus of my research to this new mystery.

Gravity is supposed to be attractive. What could be causing the cosmic speed-up? And what would this mean for the destiny of the universe? Could it be that we do not understand gravity as well as we thought? Was Perlmutter's result really correct?

The last of these questions is the only one we can answer with any certainty. A month later another group, led by Brian Schmidt of Australia, found the same amazing result: The expansion of the universe is speeding up, not slowing down. Schmidt's group also used supernovae and, coincidentally, the same digital camera. One year later, evidence from a completely different source—the cosmic microwave background—confirmed the cosmic speed-up. This magical result from nature, unlike the illusions of a magician, is not going away.

Gravity as we know it is attractive. But there is a loophole of sorts in Einstein's general theory of relativity, one that Einstein himself used. Gravity can be repulsive—not the gravity of matter, but of very unusual and exotic stuff like the energy of nothing. Einstein tried to create a static universe by balancing the repulsive gravity of a cosmological constant against the attractive gravity of matter. But when the expansion of the universe was discovered, he quickly discarded this idea, calling it "my biggest blunder."

For almost forty years physicists have known that Einstein's cosmological constant could not be discarded so easily. Like it or not, repulsive gravity can arise due to the energy of the quantum vacuum. According to classical physics, the vacuum—the empty space that exists between things—is truly empty. Quantum mechanics changed that picture radically. The quantum vacuum is alive with particles and antiparticles popping in and out of existence, living on borrowed time and energy. The existence of these virtual particles was confirmed in a beautiful experiment carried out in the late 1940s by Willis Lamb. The quantum vacuum behaves just like Einstein's discarded cosmological constant. Einstein's cosmological constant *must* be there.

The amount of repulsive gravity the quantum vacuum exerts depends upon how much the vacuum weighs. The more it weighs, the more repulsive it is (just as the gravity of the sun is a million times stronger than that of Earth, which weighs a million times less.) So in essence, by measuring the amount of cosmic speedup, Perlmutter weighed nothingness.

But there is a problem, a very big problem. The attempts of quantum theorists to calculate how much "nothing" weighs have led to the biggest embarrassment in theoretical physics: All answers are nonsensically large. This horrible discrepancy suggests to many that when we figure it all out, we will find that quantum nothingness actually weighs nothing (which would make perfect sense).

If nothing weighs nothing, something even weirder than quantum vacuum energy must be responsible for the cosmic acceleration. To give it a name, I call it dark energy, because it is more like energy than matter and you can't see it directly with telescopes. This strange dark energy accounts for two-thirds of the stuff in the universe and we don't know what it is.

So why be interested in what dark energy is? Nothing short of the fate of the universe depends upon it. If cosmic acceleration is due to quantum vacuum energy, it will continue forever, with unpleasant consequences. In 150 billion years, all but a few hundred galaxies will become too faint to see, leaving us very lonely. In fact, much worse than lonely: In an accelerating universe, cosmic energy resources are finite, making the existence of intelligent life a passing fancy. However, if the dark energy is something else, it may dissipate, resulting in a cosmic slowdown, and a much brighter cosmic future. Suggestions for that something else range from a tangled network of cosmic string filling the universe to the influence of unseen, extraspatial dimensions.

Cosmic speed is causing headaches for string theorists too (which is music to the ears of some). Since string theory is the great hope for marrying quantum mechanics to gravity and unifying all the forces, it should have something to say about the gravity of quantum energy. Not only hasn't string theory shed light on dark energy but some string theorists believe that cosmic speed-up and string theory are simply incompatible.

The years since Perlmutter's discovery have been exciting. Things have been changing fast, with new questions, more evidence for cosmic acceleration, new ideas, and mostly more confusion. We don't have a clue as to what dark energy is, but we're certain it is extremely important. We're also having the time of our lives.

Though scientists, especially theorists, hate to admit it, Nature is just smarter and more clever than we are. Just when we think we have it all figured out, bam, a big surprise. And of course, it is the certain knowledge that every once in a while Nature will surprise us again in a deep and wonderful way that keeps us going late at night. Whenever she does, we are like young children on Christmas morning.

What next? Two really good surprises in a scientific lifetime are probably too much to hope for. But anyway, I'm tied up for the foreseeable future trying to figure out what the dark energy is.

THE   PURPOSE OF SCHOOLS (MY OPINION)
IS TO PRODUCE GRADUATES WHO CAN
MANAGE* AND THRIVE IN THE WORLD
INTO WHICH THEY EMERGE

BUT THAT WORLD IS CHANGING
(IT IS NOT THE WORLD OF THE TEACHERS,
PARENTS, SCHOOL OFFICIALS, EVEN OF
PRESIDENT CLINTON ..)

* THEY MUST OF COURSE MANAGE THEIR
  OWN LIVES BUT ALSO PLAY A ROLE IN'
  DECIDING HOW THE CITY, THE STATE AND
  THE NATION WILL USE THE POTENTIAL
  OF THE AWESOME NEW TECHNOLOGIES :

  FOR THE BENEFIT OF HUMANITY
    (BUDAPEST : WCS)

        OR
  FOR GREED AND FEAR

PART 3
REFRAMING SCIENCE TEACHING

DRINK DEEP, OR TASTE NOT THE PIERIAN SPRING

Musings on the Teaching and Learning of Science

Stephen Jay Gould

Most famous quotations are fabricated; after all, who can concoct a high witticism at a moment of maximal stress in battle or just before death? A military commander will surely mutter a mundane "Oh hell, here they come," rather than the inspirational "Don't one of you fire until you see the whites of their eyes." Similarly, we know many great literary lines by a standard misquotation rather than accurate citation. Bogart never said, "Play it again, Sam," and Jesus did not proclaim that "he who lives by the sword shall die by the sword." Ironically, for this special issue on learning, the most famous of all quotations bungles the line and substitutes "knowledge" for the original. So let us restore our celebratory word to Alexander Pope's *Essay on Criticism*:

> *A little learning is a dangerous thing;*
> *Drink deep, or taste not the Pierian spring;*
> *There shallow draughts*
> *intoxicate the brain,*
> *And drinking largely sobers us again.*

I have a theory about the persistence of the standard mis-quote, "a little knowledge is a dangerous thing," a conjecture that I can support through the embarrassment of personal testimony. I think that writers resist a full and accurate citation because they do not know the meaning of the crucial second line. What the dickens is a "Pierian spring," and how can you explain the quotation if you don't know? So you extract the first line alone from false memory, and "learning" disappears.

To begin this little essay about learning in science, I vowed to find out about the Pierian spring so I could dare to quote this couplet that I have never cited for fear that someone would ask. And the answer turned out to be joyfully accessible—a two-minute exercise involving one false lead in the encyclopedia (reading two irrelevant articles about artists named Piero), followed by a good turn to the *Oxford English Dictionary*. Pieria, this venerable source tells us, is "a district in northern Thessaly, the reputed home of the muses." And Pierian therefore becomes "an epithet of the muses; hence allusively in reference to poetry and learning."

So I started musing about learning. Doesn't my little story illustrate a general case: We are afraid because we fear that something we want to learn will be hard and that we will never even figure out how to find out. And then, when we actually try, it's easy—with such joy in discovery, for there can be no greater delight than finding the definitive solution to a little puzzle. Easy, that is, so long as we have the tools at hand (not everyone has immediate access to the *Oxford English Dictionary*; more sadly, most people never learned how to use this great compendium or know that it even exists). Learning can be easy because the human mind is an intellectual sponge of astonishing porosity and voracious appetite, that is, if proper education and encouragement keep those spaces open.

A commonplace of our culture, and the complaint of teachers, holds that, of all subjects, science ranks as the most difficult to learn and therefore the scariest and least accessible of all disciplines. Science may be central to our practical lives, but its content remains mysterious to nearly all Americans, who must therefore take its benefits on faith (turn on your car or computer and pray that the thing will work) or fear its alien powers and intrusions (will my clone steal my individuality? Will greenhouse warming drown my city?). We suspect that public knowledge of science may be extraordinarily shallow, both because few people have any interest or familiarity with the subject (largely through fear or from assumptions of utter incompetence) and because those who profess concern have too superficial an understanding. Therefore, to continue with Pope's topsy-turvy metaphor, Americans shun the deep drink that sobriety requires and maintain dangerously little learning about science.

I write to argue that this common, almost mantralike, belief among educators is entirely wrong and primarily the product of a common error in the sciences of natural history (including human sociology in this case)—a false taxonomy. I believe that science is wonderfully accessible, that most people show a strong interest, and that levels of general learning stand quite high (within an admittedly anti-intellectual culture overall), but that we have mistakenly failed to include the domains of maximal public learning within the scope of science. (And, like Pope, I do distinguish learning, or visceral understanding by long effort and experience, from mere knowledge, which can be mechanically copied from a book.)

I do not, of course, hold that most people have the highly technical skills that lead to professional competence. But such is the case for any subject or craft, even in the least arcane and mathematical of the humanities. Few Americans can play the violin in a symphony orchestra, but nearly all of us can learn to appreciate the music in a seriously intellectual way. Few can read ancient Greek or medieval Italian, but all can learn to love a new translation of Homer or Dante. Similarly, few can do the mathematics of particle physics, but all can understand the basic issues behind deep questions about the ultimate nature of things and even learn the difference between a charmed quark and the newly discovered top quark.

For the false taxonomy, we don't restrict adequate knowledge of music to professional players; so why do we limit understanding of science to those who live in laboratories, twirl dials, and publish papers? Taxonomies are theories of knowledge, not objective pigeonholes, hat racks, or stamp albums with places preassigned. A false taxonomy based on a bogus theory of knowledge can lead us badly astray. When Guillaume Rondelet, in his 1555 classic on the taxonomy of fishes, began his list of categories with "flat and compressed fishes," "those that dwell among the rocks," "little fishes" (*piscicuti*), "genera of lizards," and "fishes that are almost round," he pretty much precluded any deep insight into the truly genealogical basis of historical order.

Millions of Americans love science and have learned the feel of true expertise in a chosen expression. But we do not honor these expressions by categorization within the realm of science, although we certainly should, for they encompass the chief criteria of detailed knowledge about nature and critical thinking, based on logic and experience. Consider just a small list, spanning all ages and classes and including a substantial fraction of our population. If all the following people understood that they were *doing* science, democracy would shake hands with the academy, and we might learn to harvest a deep and widespread fascination in the service of more general education. (I thank Philip Morrison, one of America's wisest scientists and humanists, for making this argument to me many years ago, thus putting my thinking on the right track.)

1. Sophisticated knowledge about underwater ecology among tropical fish enthusiasts, mainly blue-collar males and therefore mostly invisible to professional intellectuals who tend to emerge from other social classes.

2. The horticultural experience of millions of members in thousands of garden clubs, mostly tenanted by older, middle-class women.

3. The upper-class penchant for birding, safaris, and eco-tourism.

4. The intimate knowledge of local natural history among millions of hunters and fishermen.

5. The astronomical learning (and experience in fields from practical lens grinding to theoretical optics) of telescope enthusiasts, with their clubs and journals.

6. The technological intuitions of amateur car mechanics, model builders, and weekend sailors.

7. Even the statistical knowledge of good poker players and racetrack touts. (The human brain seems especially poorly built for reasoning about probability, and no greater impediment to truly scientific thinking exists. But many Americans have learned to understand probability through the ultimate challenge of the pocketbook.)

8. In my favorite and clinching example, the dinosaur lore so lovingly learned—and not merely known—by America's children. How I wish that we could quantify the mental might included in all the correct spellings of hideously complex dinosaur names among all the five-year-old children in America. Then we could truly move mountains.

Common belief is ass-backward. We think that science is intrinsically hard, scary, and arcane, and that teachers can only beat the necessary knowledge, by threat and exhortation, into a small minority blessed with inborn propensity. No. Most of us are born with a love of science (which is, after all, only a method for learning the facts and principles of the natural world surrounding us, and how can anyone fail to be stirred by such an intimate subject?). This love has to be beaten *out* of us if we are to fall by the wayside, perversely led to say that we hate or fear the subject. The love burns brightly throughout the lives of millions, who remain amateurs in the precious, literal sense of the word ("those who love") and who pursue "hobbies" in scientific fields that we falsely refuse to place within the taxonomic compass of the discipline.

And so, finally, the task of nurture and rescue goes to those who represent what I have often called the most noble word in our language, teacher. ("Parent" holds second place to "teacher" on my list; teachers come first because parents, after a certain decision, have no choice.) Rage (and scheme) against the dying of the light of childhood's fascination. And be like English literature's first instructor, the clerk of Oxenford in Chaucer's *Canterbury Tales*—the man who opened both ends of his mind and heart, for "gladly wolde he lerne, and gladly teche."

THE UNIVERSITY AS A PARTNER IN TRANSFORMING SCIENCE EDUCATION

Elnora Harcombe and Neal Lane

We are delighted to join so many colleagues and admirers in honoring Dr. Leon Lederman on the occasion of his eightieth birthday. It is accomplishment enough for an individual to advance science through path-breaking research, but in Leon's case, he then turned his attention—his creative energy and enormous powers of persuasion—toward trying to improve science and math education in this country. He has had an enormous impact, beginning with his efforts to connect K–12 education with the work of Fermi National Accelerator Laboratory, preeminent in the world of high-energy physics. Leon Lederman has always understood that young people are curious about nature, and they learn best by asking questions, so long as someone is there to encourage them and help them find the answers.

The National Commission on Mathematics and Science Teaching for the Twenty-first Century, chaired by former senator John Glenn, in its September 2000 report "Before It's Too Late" (U.S. Department of Education 2000), had this to say about the situation in the nation:

> At the daybreak of this new century and millennium . . . the future well-being of our nation and people depends not just on how well we educate our children gen-

erally, but on how well we educate them in mathematics and science specifically. . . . it is abundantly clear from the evidence already at hand that we are not doing the job that we should do. . . . [O]ur children are falling behind; they are simply not "world-class learners" when it comes to mathematics and science.

Lane was privileged to serve on the Glenn Commission and was struck by the enormity of the challenge we face as a nation and as a concerned, but still largely uninvolved, community of scientists, mathematicians, engineers, and other technical professionals.

Lederman understands this challenge well and has done something about it. Not all of us can create educational centers and academies as he did, but we can provide the leadership, sensitivity, creativity, and direction that are needed to support our teachers. The improvement of science education in the United States will depend largely on the leadership generated by our full scientific community in universities, government laboratories, and industry. It is our responsibility to prepare teachers who can nurture curiosity, guide the framing of questions, and point the path toward discovery and answers, as our students explore the world through inquiry and experimentation, as advocated by Lederman.

A MODEL

Lederman recently endorsed the book *Science Teaching/Science Learning* (Harcombe 2001) by Elnora Harcombe, which is a resource to other scientists who seek views of classroom dynamics and insights into the cultures of teachers. It is a description of ten years of challenges and successes with Houston science teachers in the Model Science Laboratory Project, established to dramatically improve the teaching of science in Houston's urban schools and led by the Rice University Center for Education, working in partnership with the Houston Independent School District (HISD). We bring this program to your attention not only because it is an example of the engaged learning philosophy promoted by Lederman, but also because this program has achieved an amazing record of teacher retention.

Over 95 percent of the teachers impacted by the Model Science Lab during the past eleven years have remained in education until death or retirement! If more teachers were retained at rates such as this, the nation would not be facing such a critical shortage of science teachers. Perhaps the main reason for these impressive numbers is the change in philosophy and outlook of the teachers who spend time in the program. Here's what one teacher had to say:

> Before attending the program, I tried to learn ways to be an effective science teacher, but I was not successful. Each school year I completed, I did not feel as though I was making a difference in my students' lives. My students always commended me on how great a teacher I was, but I did not feel that. Sometimes I found myself teaching the way I was taught, for example, lectures and bookwork. This program allowed me to blossom. I felt invigorated. I was encouraged and forced out of my comfort zone. Once I learned about inquiry-based teaching, I became excited. As the year progressed, I became more and more confident in my teaching, speaking in front of adults, creating original work/curriculum, and writing. (CJ)

TIME, SCIENCE, AND STUDENT FOCUS

The Model Science Laboratory Project is one approach that combines the various features of science education that have been shown to be important—time for teachers to reflect and grow, science content enrichment through inquiry and interaction with scientists, and opportunity for immediate application of insights to a class of students while monitoring instructional effectiveness with respect to deep student understanding. The format is a structured and focused sabbatical residency program. Each year, eight middle school teachers are released from their normal classroom duties to join the Model Science Lab that operates in an urban middle school in the center of Houston. A typical two-week schedule is shown in Table 1. On a teaching day, each resident teacher instructs one class of twenty-four middle school students with a team teacher for ninety minutes, and uses the rest of the day to analyze and debrief on the class and plan for the next class in detail using multiple resources. On alternate days, the resident teachers attend special Rice University classes offered on-site at the middle school. The teachers learn science concepts in the way they are encouraged to teach, by way of inquiry. In addition, they gain professional extension through visits to research laboratories, field trips, special lectures, conferences, and observing other teachers. The eight teachers form a strong professional supportive network as they grapple together with integrating all of their new information and past experiences to create an effective learning environment for urban youth. The Model Lab provides teachers a safe place to try new ideas, to share with peers, to practice technology as a tool in instruction, and to experience leadership by making presentations at professional workshops. The focus always returns to promoting student understanding of science concepts rather than memorized words.

At the end of the residency year, the teachers are required to return to their home schools for at least one year. They return to the Model Lab for voluntary monthly meetings and to present workshops. Model Lab staff members visit the classrooms of past resident teachers and continue to encourage them to further their academic endeavors and leadership development. The peer network helps to alleviate the sense of isolation.

RESULTS: STUDENT PERFORMANCE

The Model Lab program not only caused a transformation in teachers' perspectives but also impacted the students in their classrooms. Primarily, students became more engaged in their learning. The effectiveness of inquiry instruction adopted by most of the Model Lab teachers is not well measured by standardized tests. Nevertheless, when students taught by Model Lab graduates were compared with students taught by control teachers in the same schools, in a pre-test/post-test situation, there was signifi-

TABLE 1. TYPICAL ACTIVITY SCHEDULE FOR EIGHT RESIDENT TEACHERS IN MODEL SCIENCE LABORATORY

Week 1

	Monday	Tuesday	Wednesday	Thursday	Friday
a	Teachers 1 & 2 teach class of students.	Science content class for teachers.	Teachers 1 & 2 teach class of students.	Pedagogy class for teachers.	Teachers 1 & 2 teach class of students.
m	Teachers 3 & 4 teach class of students.		Teachers 3 & 4 teach class of students.		Teachers 3 & 4 teach class of students.
p	Teachers 5 & 6 teach class of students.	Teachers observe and analyze an expert teacher in HISD.	Teachers 5 & 6 teach class of students.	Guest scientist presents research.	Teachers 5 & 6 teach class of students.
m	Teachers 7 & 8 teach class of students		Teachers 7 & 8 teach class of students.		Teachers 7 & 8 teach class of students.

Week 2

	Monday	Tuesday	Wednesday	Thursday	Friday
a	Science content class for teachers.	Teachers 1 & 2 teach class of students.	Pedagogy class for teachers.	Teachers 1 & 2 teach class of students.	Teachers attend training in national programs, such as SEPUP, Fast Plants, FOSS, WET; or take field trip; or present workshop; or go to conference.
m		Teachers 3 & 4 teach class of students.		Teachers 3 & 4 teach class of students.	
p	Interview student for case study; or writing class.	Teachers 5 & 6 teach class of students.	Technology class; or interact with home school.	Teachers 5 & 6 teach class of students.	
m		Teachers 7 & 8 teach class of students.		Teachers 7 & 8 teach class of students.	

cantly (p<0.01) greater progress displayed by students of Model Lab teachers. This greater improvement of scores occurred on a modified National Assessment of Educational Progress (NAEP) science content test and also on the mathematics portion of the Texas Assessment of Academic Skills (TAAS) test (there is no yearly science TAAS).

RESULTS: TEACHER RETENTION

Retention may be the best numerical indicator of the effectiveness of the program as it reflects the rededication of the teachers. As stated previously, of the eighty-two teachers who have been in the residency over the past eleven years, 95 percent remain in education. This retention is remarkable in a school district that has the typical urban turnover of nearly one-seventh of the teachers each year. The Model Lab teachers have remained committed to urban children, with 74 percent remaining in HISD. Those who left HISD usually followed spouses to other urban settings, and only 9 percent moved out to suburban schools. Furthermore, the Model Lab teachers have generally remained in instruction, with 74 percent of the original number still in classrooms and only 17 percent moving into administration. The teachers have created their own leadership opportunities in mentoring other teachers; by assisting in curriculum development; as officers in professional organizations, and through presentations to local, state, national, and international audiences.

MODEL SCIENCE LABORATORY

A cornerstone of the Model Science Laboratory project is to provide science-learning opportunities for teachers where they can struggle with nature and ideas, just as research scientists do, and experience the thrill of discovery and generating insights into science concepts. They experience the kind of *ah-ha* achievement that comes with true learning, and develop a renewed (or new) passion for inquiry that can sustain them well into the future. When teachers become engaged in their own active learning, they usually become eager to share the excitement with their students, and this transforms their teaching. Listen to this Model Lab teacher:

> The program modeled for me the idea that telling someone something is the least effective way to make a change. I had to discover for myself what changes needed to be made in my teaching in order to have the motivation and drive to recon-struct it.
>
> What I have discovered about effective teaching is simple; it focuses on my definition of understanding. My early definition was related to written and verbal recall of information pre-sented, but this has changed. Now, I feel that learning encom-passes understanding a concept well enough to be able to demonstrate, explain and apply this knowledge. For the teacher, it means comprehending how the learner internalizes and forms these concepts. This knowledge has been the catalyst for changing the way I teach. The director supported me to take risks and grow. I am a learner and a teacher who has only begun to make changes. (NG)

The key points in this testimony are (1) telling something is the least effective way to teach and (2) learning encompasses understanding a concept well enough to use it. NG was a very per-ceptive teacher who cared passionately about her students. She returned to her classroom after the Model Lab experience and was so effective with her students that her entire school began fol-lowing her example and even rearranged budgets and schedules so that NG could assist peers in their classrooms.

The science content aspect of the Model Laboratory project is balanced with a focus on how young people learn. There is an extensive literature now about inquiry, constructivism, and teaching for understanding (Cohen, McLaughlin, and Talbert 1993; Meier 1995; National Research Council 1996; Watson and Konicek 1990; Wiske 1997). *The Private Universe* series of video-tapes[1] has been particularly powerful in demonstrating the existence and persistence of students' naïve conceptions. Yet reading, hearing, or even seeing teaching modifications leaves one a long way from being able to implement such approaches in a classroom. Reforming one's behavior is even less likely to happen than learning science concepts if one is only told about it. Teachers need the opportunity to practice new teaching methods and convince themselves that their new approaches really are more effective with students before they are willing to abandon their long-established routines.

The two most unique features of the Model Science Lab project are the creation of a safe environment to practice new approaches to teaching while examining their effectiveness and the gift of time to analyze and reflect. Teachers almost never have time to think calmly, time to learn in depth, time for curiosity, or time to integrate. If we in our rapidly changing global society expect our teachers to stimulate students to become lifelong learners, then we need to provide teachers the opportunity to nourish their own intellectual life. Teachers need time to think quietly. They also need encouragement to be creative, to apply ideas, to devise new approaches. This is clearly identified by a Model Lab teacher who writes about the importance of the reflection time:

> My year as a resident at the Rice Model Science Lab forced me to take a critical look at the way I teach, and more importantly, to determine if my lessons were effective in improving the understanding of the students. It gave me the chance to examine my beliefs about teaching style, to compare various approaches of presenting concepts to the students, and to record both my successes and failures in the classroom. Writing my journal forced me to be honest with myself and to question the validity of what I did in my classroom.
>
> I'm sure the full impact of this year has not yet been felt. Each year I teach after this will be an attempt to create an ideal learning atmosphere using the methods and ideas explored at the Model Lab. I am truly excited and I'm sure my students will be too. (NK)

This enthusiasm for effective teaching contrasts strongly with the burned-out, ready-to-quit attitude this veteran of twenty-one years of teaching was feeling when he entered our program. Notice how he echoes NG's conviction that the insights gained in the Model Lab are just the beginning of a challenging new journey that will involve continued learning and evolving instructional innovations. This conviction about a developing future is one basis for the rededication to teaching found in the participants of the Model Lab that leads to the amazing retention rate.

The intervention of the Model Lab has been successful with veteran teachers, as shown by the previous quotes. It has also impacted younger teachers such as this one:

> Being a resident teacher has made me recognize that teaching is not only a career, but also a profession. I have changed my belief about teaching. Once believing that teaching was a mediocre job with a two month paid summer vacation, I now believe that teaching is the founding profession of ALL professions. Teachers develop the knowledge of nurses, doctors, policemen, trainers, household managers, lawyers etc. (OW)

The prior life goals of this bright third-year teacher were aimed outside of education. She entered the Model Lab primarily at the insistence of her principal, but she gradually developed her own personal commitment to the profession.

ROLE OF THE UNIVERSITY

Scientists and university administrators have a crucial lead role in establishing programs to impact science teachers. At Rice, the Model Lab concept was developed in correlation with the expressed vision of former President George Rupp that the university should extend itself "beyond the hedges" of the campus by contributing to the greater community in Houston, especially in the area of science education. Rice's involvement remains strong today. Support at the top level of the institution can create the expectation that faculty involvement in public K–12 education will be looked upon favorably, in stark contrast to the common perception that faculty interaction with local schools is a distraction that can compete dangerously with one's productivity in research.

The support by the Rice administration went beyond setting new institutional expectations. It also included personal involvement in soliciting funds for educational projects. Leadership at the top was the catalyst that created the conditions in which individuals from the university, community, and schools who were concerned about public education could come together to discuss the issues, brainstorm solutions, and propose innovative approaches.

We can no longer afford to pretend that merely offering teachers additional university courses in science will enable them to stimulate young minds. Some teachers have had extensive science preparation yet still reported sentiments similar to this teacher's:

> Before entering the program, I had experience in biochemistry and genetics research, geophysical data processing, and a master's degree in education, but this background had not transferred to effective teaching. I taught the way I had been taught and conformed to a traditional educational system focused on a well-defined curriculum. Though I did not feel as if I was making a difference in the education of my students, I also lacked a model or a clear reason for changing my philosophy and strategies. (NG)

Her statement "I taught the way I had been taught" may be the most instructive for us. We, the faculty of the university, create the science-learning environment for prospective K–12 teachers who will use our model when they lead their own classrooms. If we primarily give formal lectures and focus our courses on a set of established concepts and a fact-based compendium to be memorized, then how can we expect teachers to spark the curiosity and questioning of younger students? Even more serious is whether teachers will leave the university with any experience in making sense of information, synthesizing concepts into an explanation, or pulling together information into a personal understanding—all the elements of scientific research—and thus be able to guide students in the same intellectual skills. Lillian McDermott has been vocal about the discrepancy between theoretical physics taught in college and the concrete application of physics concepts that high school students need to understand (McDermott 1996).[2] We are asking the teachers to perform conversions that many of us might find difficult even with our deeper understanding of the concepts. Teachers are expected to teach in a more engaging style, incorporate integration of ideas, and transform theory into concrete applications, as well as all the other numerous school requirements. Eleanor Duckworth of Harvard asks, "Can we expect our teachers to teach in a way in which they have never learned?" (Duckworth 1987). Teachers tend to instruct science in the way they have seen us model for them in multiple semesters rather than in the way they see in one semester of science methods in the education department.

THE FUTURE

We believe that if more university programs were tuned to the teachers' true needs, the nation would not be facing a crisis in science and math education. Teachers have demonstrated that they are willing, indeed eager, to do whatever is necessary to become the best they can be. But, too often, they leave the university lacking the knowledge and tools they need. When universities begin to better appreciate the challenges faced by teachers, then perhaps the teachers will graduate better prepared to teach science enthusiastically and effectively.

The enterprise will require the combined efforts of university administrators, scientists, educators, community leaders, and school district personnel, with the initial impetus provided by the leadership of universities and scientists on their faculty. It is well worth our effort. The children in our schools are the undergraduates and graduate students in our universities in the very near future. But, more than that, these children are our nation's future.[3]

NOTES

1. The Private Universe Project aired a series of nine teleconferences from October 13, 1994, through December 15, 1994, through the Harvard-Smithsonian Center for Astrophysics, 60 Garden Street, Cambridge, MA 02138; Science Media Group of the Science Education Department. The series is now available in other formats from the Annenberg/CPB Math & Science Collection.

2. L. C. McDermott, "Physics Education Research: The Key to Student Learning," Oersted Award Lecture, American Association of Physics Teachers (Winter Conference, 2001).

3. The authors gratefully acknowledge the resident teachers for permission to use excerpts from their professional portfolios.

REFERENCES

Cohen, D. K., M. W. McLaughlin, and J. E. Talbert, eds. *Teaching for Understanding: Challenges for Policy and Practice*. San Francisco: Jossey-Bass Publishers, 1993.

Duckworth, E. *"The Having of Wonderful Ideas" and Other Essays on Teaching and Learning*. New York: Teachers College Press, 1987.

Harcombe, E. S. *Science Teaching/Science Learning: Constructivist Learning in Urban Classrooms*. New York: Teachers College Press, 2001.

McDermott, L. C. *Physics by Inquiry*. New York: John Wiley & Sons, Inc., 1996.

Meier, D. *The Power of Their Ideas: Lessons for America from a Small School in Harlem*. Boston: Beacon Press, 1995.

National Research Council. *National Science Education Standards*. Washington, D.C.: National Academy Press, 1996.

U.S. Department of Education, The National Commission on Mathematics and Science Teaching for the Twenty-first Century. *Before It's Too Late: A Report to the Nation*. Washington D.C.: U.S. Government Printing Office, 2000.

Watson, B., and R. Konicek. "Teaching for Conceptual Change: Confronting Children's Experience." *Phi Delta Kappan* 71 (1990): 680–85.

Wiske, M. S., ed. *Teaching for Understanding: Linking Research with Practice.* San Francisco: Jossey-Bass Publishers, 1997.

THE IMPOSSIBLE TAKES A LITTLE LONGER

Dudley Herschbach

My father had a favorite saying: "The difficult we do immediately, the impossible takes a little longer." The second phrase appealed to me as a kid, I suppose because of its whimsical incongruity and heroic ring. Decades later, the appeal is stronger still. Again and again, I've seen intrepid, persevering innovators achieve wonderful things that had been regarded as "impossible." Leon Lederman is such a dauntless innovator, both as a scientist and educator. It is a special pleasure to offer homage to him and his efforts to enhance science education, which he has long pursued with inspiring verve, devotion, and joy.

This essay deals chiefly with observations and proposals pertaining to high school science teaching and learning. These stem from opportunities to talk with many students and teachers at high school science fairs and summer programs, as well as my experience teaching freshman chemistry at Harvard University over the past twenty years. First, however, I recount a story of an "impossible" educational triumph. Beyond its explicit message regarding educational strategy, this story implicitly conveys key aspects of the scientific enterprise, including the practical value of curiosity-driven research and the kinship of science and the humanities as liberal arts.

A BRILLIANT EDUCATIONAL EXPERIMENT

The story (Bruce 1990; MacKay 1997) involves Alexander Graham Bell, Helen Keller, and Keller's teacher, Annie Sullivan. Bell patented the telephone in 1876 at the age of twenty-nine, and thereafter devoted many years to pursuit of other inventions. But Bell regarded his life's work as the education of the deaf. His mother and his wife were deaf, his father and grandfather were teachers of speech and elocution. His family had emigrated from Scotland after both of his older brothers had died of tuberculosis, and settled near Toronto. Bell, in 1871, began teaching the deaf in Boston and the next year founded his own school there. It was an effort to develop a device to help his deaf students distinguish between the letters p and b that led to his invention of the telephone. As his biographers emphasize, all the technologies needed to produce the telephone existed by 1872, and several expert electricians were pursuing kindred devices. In later years, Bell liked to say that he would not have conceived the principle of the telephone if he had known more about electricity but less about the mechanics of human speech.

Helen Keller, deaf and blind since an illness at nineteen months, was six when brought to Bell by her parents to ask his opinion as to whether she could be educated. Bell instituted the arrangements that led to Annie Sullivan becoming Helen's teacher. Annie was then twenty years old, still suffering the effects of trachoma, which had made her temporarily blind for a few months (and would make her permanently blind in her old age). Annie had no teaching experience and not much to guide her except her good sense and acute sensitivity (Lash 1997). Yet only a month after she joined Helen at her parents' home in Tuscumbia, Alabama, in March 1887, came the marvelous moment when Helen discovered things had names, as Annie held her hand in a stream of water. Within three months, Helen was writing brief letters. Within three years, she had an astonishing command of idiomatic English.

This, and what followed in Helen's life, was justly hailed as a miracle. But Bell had a different view. Especially in the early years, and for decades after, he helped and supported Helen and Annie in many ways. Helen dedicated her first autobiography to him. Annie wrote of the "immense advantage" it was to have his

Helen Keller and Annie Sullivan with Alexander Graham Bell, July 1894, when Helen was fourteen years old, Annie twenty-eight, Bell forty-seven. With her left hand, Helen is reading Annie's lips, with her right hand, communicating with Bell.

advice, and her gratitude for his "happy way of making people feel pleased with themselves." Bell insisted that Helen's mastery of idiomatic English was "not a case of supernatural acquirement [but] . . . a question of instruction, . . . a brilliant experiment" by Annie (Bruce 1990). He concluded that the key was Annie's "constant spelling of natural, idiomatic English into Helen's hand without stopping to explain unfamiliar words and constructions, and her encouragement of Helen's reading book after book in Braille . . . with a similar reliance on context to explain new language." Bell stressed that this was equivalent to the way a hearing and sighted child learns. Annie's own description of her teaching confirms that she purposely *never explained anything, unless Helen asked a question.* Thereby Annie helped Helen to discover and actively exercise her ability to discern clues and context and to learn on her own.

SCIENCE AS DECIPHERING NATURE'S LANGUAGE

The reticent teaching style of Annie Sullivan is intrinsic to frontier research in science. Nature is a reticent teacher, who speaks to us abundantly but in many alien tongues. She does not offer explanations; it's up to us to ask probing questions and to generate our own understanding. In frontier research, we try to discover or add to knowledge of the vocabulary and grammar of some strange dialect. To the extent we succeed, we gain the ability to decipher many messages that Nature has left for us, blithely or coyly. No matter how much human effort and money we might devote to solving a practical problem in science or technology, failure is inevitable unless we can read the answers that Nature is willing to give us. That is why basic curiosity-driven research is an essential and practical investment, and why its most important yield are ideas and understanding (Herschbach 1996).

We are all born blind and deaf to much of Nature's language, and it takes persistent groping and guessing to learn something of it. In my classes, I like to emphasize this. I ask how many students have already studied a foreign language, and recommend that they approach science the same way: "Once you get it in your ear, it gets easier and easier; otherwise, harder and harder!" Actual counts have shown that in introductory science textbooks for high school or college, the number of new words or ordinary words used with special meanings exceed the vocabulary of a typical one-year language course. Likewise, the array of interlocking concepts met in a science course functions much like grammatical rules. In my freshman chemistry course, I point out that our triad of major topics has some resemblance to the curriculum that Harvard had in the seventeenth and eighteenth centuries, chiefly Latin, Greek, and Hebrew. The universal scope and rigor of thermodynamics resembles Latin; the elegance and poetic character of quantum theory underlying electronic and molecular structure resembles Greek; the pragmatic and forthright style of chemical kinetics resembles Hebrew.

SCIENCE AMONG THE LIBERAL ARTS

The "impossible" empowerment of Helen Keller by language exemplifies in a compelling way the highest aim of a liberal arts education: to instill the *habit of self-generated questioning and thinking*, of actively scrutinizing evidence and puzzling out answers. That is also the essence of a genuine scientific literacy. It accords with a favorite definition equally applicable to science and the humanities, "Education is what's left after all you've learned has been forgotten." This defines the aim to be understanding rather than ritualistic training; cultural perspective and self-reliant thinking rather than conventional knowing. The "what's left" aspects of science and mathematics offer much that transcends any technical particulars. For both novice scientists and students destined for other careers, teachers should emphasize the human adventure of intellectual exploration, replete with foibles and failures, but ultimately achieving wondrous insights. This is important not merely as seasoning for hearty servings of lectures, homework problems, and laboratory work but to nurture perspectives akin to the liberal arts.

A fervent appeal to cultivate common ground, shared by science and liberal arts, was made by Isidor Rabi, one of Leon Lederman's mentors, in a lecture I heard in 1955 as a beginning graduate student:

> To my mind, the value of science or the humanities lies not in the subject matter alone, or even in greater part. It lies chiefly in the spirit and living tradition in which these different disciplines are pursued. . . . Our problem is to blend these two traditions. . . . The greatest difficulty which stands in the way is communication. The nonscientist cannot listen to the scientist with pleasure and understanding.
>
> Only by the fusion of science and the humanities can we hope to reach the wisdom appropriate to our day and generation. The scientists must learn to teach science in the spirit of wisdom and in the light of the history of human thought and human effort, rather than as the geography of a universe uninhabited by mankind.

Rabi later discussed these concerns (Rigden 1987) with C. P. Snow, who developed them further in his famous *Two Cultures*.

I've pursued a liberal science approach in my freshman course (Herschbach, "Teaching Chemistry" 1996) to bring out "what's left." This is not only in response to Rabi's appeal, but because I feel the chief aim should be to entice students to take ownership of scientific ideas. That is fostered by presenting science in a more humanistic mode. I typically introduce each major topic with a story, usually having the character of a parable. Many of the parables deal with historical episodes or current research discoveries; some are fictional, designed to indulge in whimsical fun while delivering a serious message. Often the stories emphasize the role of analogy and guesswork or show how error and failure are prevalent in science but can lead to progress if "wrong in an interesting way." Usually the parable also poses questions for students to work out.

For example, when discussing the gas laws, I ask students to consider a fancied task that might have been asked of Hercules (Herschbach 1999). What if that mighty hero, after completing his legendary twelve labors, had been asked to weigh the Earth's atmosphere? The students discover that just a couple of elementary ideas suffice to estimate this, and are impressed by the magnitude (six billion megatons). Then we discuss a moral. If Hercules had failed in this "thirteenth labor," it would testify that even superhuman strength and courage cannot prevail when what is needed is an intellectual concept. Would not such a lesson improve on that conveyed by the ancient myth?

WHAT MAKES "IMPOSSIBLE" POSSIBLE?

Science enjoys a tremendous advantage: The goal—call it truth or understanding—*waits patiently to be discovered*. Thus, ordinary human talent, given sustained effort and freedom in the pursuit, can achieve marvelous advances. Far more formidable are enterprises such as business or politics; there the objectives may shift kaleidoscopically, so a brilliant move often proves a fiasco rather than a triumph because it comes a little too soon or too late. The patience of scientific truth has another important consequence. Frequently, what might appear as the most promising approach does not pan out; there are unanticipated roadblocks. Then it is vital to have some maverick scientists willing to explore unorthodox paths, perhaps straying far from the route favored by consensus. In science, it is not even desirable, much less necessary or possible, to be right at each step. At the frontier, scientists are heading in wrong directions much of the time, optimistically looking for new perspectives.

Science as encountered in typical high school or college introductory courses is strikingly at odds with this adventurous character of research as well as the spirit of liberal science. Such courses too often come across to students as a frozen body of dogma. The questions and problems seem to have only one right answer, to be found by some canonical procedure. The student who does not easily grasp the "right" way, or finds it uncongenial, is likely to become alienated. There seems to be very little scope for a personal, innovative experience.

Nothing could be further from what actual frontier research is like. At the outset, nobody knows the "right" answer, often not even the right question or approach. So the focus is on asking an interesting question or casting the familiar in a new light. In my freshman chemistry course, I explain this to the students and ask them to write poems about major concepts, because that is much more like doing real science than the usual textbook exercises. I also show them quite a few poems that pertain to science, often without intending to. For instance, here is a quatrain by Jan Skacel, a Czech poet (Kundera 1966):

> Poets don't invent poems;
> The poem is somewhere behind;
> It's been there a long, long time.
> The poet merely discovers it.

The social organization of science also has a major role in fostering "impossible" achievements. An especially persuasive case for this was given by Michael Polanyi in his classic essay *The Republic of Science* (Polyani 1962). He contrasted the hierarchical systems customary in practical affairs with the chaotic freedom of science. In the hierarchical organizations, units are directed by a chain of officers who report up the chain and assign tasks. Science proceeds very differently and much more efficiently. Each unit is on its own, free to pursue its own interests. Nonetheless, the independent units are coordinated by an *invisible hand*, because each has the opportunity to observe and apply the results found by the others. This creates a community of scientists that amplifies individual initiatives.

Again, there is an ironic contrast between such intrinsic cooperation and the artificial competition among students that is imposed in typical courses. Instead, I use an absolute grading scale, so students compete against my standard, not each other. Students thereby are encouraged to help each other in study groups as well as in some work that is done in teams. Conducting class discussions in which groups of three or four students consult with each other before proposing or endorsing a solution to a problem usually proves instructive and lively fun. Moreover, that mode turns out to encourage students to formulate and defend guesses. That is fundamental for science, but it is actually inhibited by the usual academic rituals. I like to tell my classes: "Not so many years from now, most of you will be considered expert in something. Then you will find that clients often come to ask your opinion, not because of what you know, but because they think as an expert you can guess better than they can."

EMPOWERING STUDENTS AS TEACHERS

The National Science Foundation is required to carry out, at intervals of a few years, a survey comparing the performance of U.S. students in science and mathematics with those of other nations. For more than two decades, the results have repeatedly shown a dismaying pattern: U.S. fourth-graders perform well above the world average, eighth-graders about average, but twelfth-graders far below average. A major contributing factor is the widely deplored shortage of qualified high school teachers of science and mathematics (Gregorian 2001). Many efforts have been undertaken to recruit science teachers and enhance their training. Among these are the excellent "Teach for America" program (Shapiro 1993) launched in 1988 by Wendy Kopp when she was a Princeton undergraduate, and a markedly successful Teachers' Academy in Chicago (Lederman 1998; Sparks and Hirsh 1997), cochaired by Leon Lederman. Yet, nationwide, the problem remains daunting.

I do not believe we can provide an adequate corps of science teachers in the foreseeable future. However, I am convinced that this gap could be significantly offset by empowering able students as teachers to a much greater extent than occurs today. This conviction has two distinct sources: recent science fairs and reflections on my own high school days.

More than a decade ago, I was recruited by Glenn Seaborg to join the board of a small nonprofit outfit, Science Service, in Washington, D.C. As well as publishing the weekly *Science News*, written for laypeople, it is the premier sponsor of high school science fairs. For sixty years it has conducted the Science Talent Search (STS) originally sponsored by Westinghouse and recently by Intel. For nearly as long it has run the International Science and Engineering Fair (ISEF), now also sponsored by Intel. This brings together more than 1,200 students from 50 countries (95 percent from the United States), winners of hundreds of local, state, and regional fairs in which about a million other students took part. A host of adult volunteers, among them many parents, teachers, and scientists—including Leon Lederman—gladly contribute to the infrastructure of these fairs.

The STS and ISEF finalists are fine ambassadors for science. Many of them have done impressive, original projects, often facilitated by summer research opportunities or links with university laboratories found via Web searches. Without prompting, these students frequently express concerns about high school science courses, at their own schools and others. Quite a few STS and ISEF alumni attend Harvard. Five years ago, they launched a *Journal of Undergraduate Science*, with alternate issues comprised of articles that had originally been submitted for STS or ISEF projects. Those special issues were sent to 1,000 high schools, to show students and teachers what good projects are like. Recognizing the need, the editors provided for each article in those issues an extensive supplementary guide, explaining pertinent vocabulary, background, and concepts in a way accessible to typical high school students and teachers. This effort, entirely a student initiative, could be a harbinger. Modest funding and provision of Web links would enable a network to be created with which college science students can provide significant help to both students and teachers at their former high schools.

The other source for my optimistic conviction is my own high school experience. It was five decades ago, in a rural area; not many of the students expected to go on to college. Most of the science and math courses were taught by current or former athletic coaches, often the case even now (Rogers 2001). On my first day of high school, the first class was algebra. The teacher began by saying, "I don't know much about algebra!" Within a week or so, a few students indeed were well ahead of the teacher. That was not a problem for him; as a former army officer, he viewed his job as making sure the troops measured up to standards. He got the capable kids to explain things to others and to him. The result was a free-flowing discussion, in which nobody was inhibited in asking questions. Much the same happened in other courses. In chemistry we had a teacher with admirable command of the subject, but he, too, put great responsibility on the most able students to lead the class. Although we had no opportunity to take part in science fairs, we were challenged by our teachers to become genuine partners in the educational enterprise. The value of "making a virtue of necessity" seems just as compelling today (Lederman 2001).

A striking demonstration of what gifted high school students can contribute to enhancing science literacy comes from another Lederman project. Fifteen students from the Illinois Mathematics and Science Academy undertook to write biographies of leading American scientists. Each young author chose a favorite scientist to study and interview, and produced a lively account of his or her life and work, written for middle school and high school students. The result is a remarkable book (Lederman and Scheppler 2001) of portraits of inspiring scientists, doubly engaging as a legacy of roots and wings insightfully received by a new generation.

A NOBEL BENEDICTION

Leon Lederman took part in the Nobel Prize Centennial festivities held in December 2001, in Stockholm, as a member of a panel discussing the challenges looming in the coming century. Witnessing Lederman and other laureates acknowledging the "impossible" problems ahead as well as new opportunities for science in the service of humanity, I was prompted to reflect on an aspect of the prizes that is seldom noted.

Images of the gold Nobel medal are often displayed, but usually only the side depicting the profile of Alfred Nobel. The reverse face of the medal deserves attention as it intends to convey the significance of the awards (Lagerqvist 2001). The medals for

Nobel medal: Reverse face of the Nobel medal for physics or chemistry, depicting Scienta gazing upon the face of her teacher, Natura.

physics, chemistry, physiology or medicine, and literature were all designed by a Swedish artist, Erik Lindberg, in 1902. The reverse face of each of the four medals bears along its upper border a Latin inscription: "Inventas vitam iuvat excoluisee per artes." In English this may be approximated as "It is a pleasure to have brought cultivation to life through the discovered arts." This inscription, adopted by the Nobel Foundation for the science prizes as well as the literature prize, emphasizes the cultural kinship of science and the humanities.[1]

For the physics and chemistry medals, the reverse face depicts two elegant women in diaphanous gowns, both perched among billowing clouds, one erect, the other kneeling. The erect figure, designated "Natura," holds in her right hand a cornucopia. The kneeling figure, "Scienta," wears a laurel crown and has in her left hand a scroll. She reaches up with her right hand to unveil the face of Nature, which she beholds intently. The unveiling motif, familiar in classical antiquity, aptly represents the aim of scientific research. It pertains just as well to science education. For me, this reaches beyond metaphor. Lindberg's earnest and lovely figures readily morph into another immortal pair: teacher and student, Annie Sullivan and Helen Keller.

NOTE

1. Langerqvist states that the inscription is "a revision of Virgil's (Aeneid, 6:663) *inventas aut qui vitam excoluere per artes.*" The English translation that I have quoted was kindly provided by Dr. Richard F. Thomas, Professor of Greek and Latin at Harvard University. He also pointed out that the Latin *artes* itself is a broad term that can include science and technology as well as arts and crafts.

REFERENCES

Bruce, Robert V. *Bell: Alexander Graham Bell and the Conquest of Solitude.* Ithaca, N.Y.: Cornell University Press, 1990.

Gregorian, Vartan. "Teacher Education Must Become Colleges' Central Preoccupation." *The Chronicle Review, The Chronicle of Higher Education* (August 17, 2001), pp. B7–B8.

Herschbach, Dudley. "Imaginary Gardens with Real Toads." *Annals of the New York Academy of Sciences* 775 (1996): 11.

Herschbach, Dudley. "Teaching Chemistry as a Liberal Art." *Liberal Education* 82 (1996): 1–9.

Herschbach, Dudley. "The Thirteenth Labor of Hercules." In *The Thirteenth Labor: Improving Science Education*, eds. E. J. Chaisson and T-C. Kim, pp. 61–70. Amsterdam: Gordon and Breach Publishers, 1999.

Kundera, M. *The Art of the Novel.* New York: Grove Press, 1966, p. 99.

Lash, Joseph P. *Helen and Teacher.* New York: Addison-Wesley Publishers, 1997.

Lagerqvist, Lars O. *Nobel Medals.* Stockholm: The Royal Coin Cabinet, 2001, pp. 20–21.

Lederman, Leon. "Lessons Learned: The Teachers Academy for Mathematics and Science." *Phi Delta Kappan* (October 1998): 158.

———. "Revolution in Science Education: Put Physics First!" *Physics Today* 54 (September 2001): 11–12.

Lederman, Leon M., and Judith A. Scheppler, eds. *Portraits of Great American Scientists.* Amherst, N.Y.: Prometheus Books, 2001.

Mackay, James. *Sounds Out of Silence.* Edinburgh: Mainstream Publishing, 1997.

Polanyi, Michael. "The Republic of Science: Its Political and Economic Theory." *Minerva* 1 (1962): 54.

Rigden John S. *Rabi, Scientist and Citizen.* New York: Basic Books, 1987, pp. 256–57.

Rogers, T. K. "The View of Physics from High School." *APS News* 10 (August/September 2001): 8.

Shapiro, Michael. *Who Will Teach for America?* Washington, D.C.: Drew Fairchild, Inc., 1993.

Sparks, Dennis, and Stephanie Hirsh. *The New Vision for Staff Development.* Alexandria, Va.: Association for Supervision and Curriculum Development, 1997, p. 49.

Selling Physics to Unwilling Buyers

Physics Fact and Fiction

Lawrence M. Krauss

Many physicists have had the following experience. You meet someone at a party, and they ask you what you do. You tell them you are a physicist. Quickly, they change the topic. But if you ask them if they are interested in black holes, warp drives, or time travel, then they are fascinated.

Most people think that they have little interest in physics, and yet at the same time they are remarkably interested in many of the things that physics deals with. This dichotomy suggests that we have done a poor job of relating physics to the nonphysicist and that the natural way to get people motivated to learn about our field would be to stress the connection between physics and their own interests.

I sometimes have the opportunity to lecture to teachers about teaching and, when I do so, I usually point out that the biggest mistake any teacher can make is to assume that the students are interested in what you have to say. Instead, you have to be prepared to convince them to be interested, and you cannot expect them to come to you. Rather, you must reach out to where they are. I think that this maxim applies far more broadly than just to classroom

"Selling Physics to Unwilling Buyers: Physics Fact and Fiction," by Lawrence Krauss, first appeared in *Physics World* no. 11, 7 (July 1998): 13–14.

teaching, but at any rate it certainly applies to public education. Motivation is far more important than clarity, initially at least.

This is all to preface why I—someone who likes to think of himself as a reasonably respectable physicist—found myself writing and lecturing on the physics of *Star Trek*. After all, *Star Trek*, as I have to remind many individuals dressed in uniform at my lectures, is science fiction. The show makes no pretense to describe reality, nor to need it. As Gene Roddenberry, the show's creator, said, the Starship *Enterprise* is primarily a vehicle for drama. The science is thrown in and arbitrarily bent to fit the needs of the plot—not vice versa.

Nevertheless, *Star Trek* has captured the public's imagination. For example, when the Air and Space Museum in Washington, D.C., had an exhibit of the *Enterprise*, it was the most popular exhibit in the entire history of the museum—far more popular than any real spacecraft that had actually traveled in outer space!

What better way could there be, it seemed to me, to try and reach people than to use an icon of popular culture? As I pondered the issue, I recognized that the series touches on a range of diverse physical phenomena in one way or another. Moreover, I decided that one of the reasons why the series has been so popular with the viewing public over the past thirty years is that it is about *possibilities*. Surely this is why most physicists do physics? After all, they simply want to know what is possible in the universe. Thus the idea of using the *Star Trek* setting as a laboratory in which to explore the physics of the real universe began to become more and more natural in my mind.

GROUND RULES

This is not to say that I did not have misgivings about the whole effort. Since it goes without saying that much of science fiction— and indeed a great deal of *Star Trek*—involves scientific nonsense, does it diminish the real world of physics to delve into such fantasy worlds? Moreover, what is the point of debunking a fictional universe? And indeed, how would my colleagues view the effort, and how would the fans of *Star Trek* react to what I was doing?

I decided early on that there had to be several ground rules if this effort were not to revert either into an apologia for the indiscretions of *Star Trek* writers, or into a nit-picking diatribe that would be of interest to no one. First, no matter how much it hurt, if something was impossible, I would say so. Second, rather than dwell on these impossibilities, if something in the fiction were impossible, I would find something in the real world to relate it to that might not be.

This whole enterprise has reinforced my conviction that truth is indeed stranger than fiction, and I think that people are most taken by this when examples are thrust in their faces. Indeed, in one of my favorite reviews of my most recent book, *Beyond Star Trek*, which appeared in the U.S. magazine *Publishers Weekly*, the reviewer—much to his surprise apparently—acknowledged that scientific phenomena are often far more fascinating than fictional ones. I find no better justification for using science fiction as a way to teach science than this demonstration that it can convince people that the real world is fascinating.

I have yet to talk to an audience of laypeople about, say, solar neutrino detection in the context of *Star Trek* bloopers about neutrinos, without hearing titters erupt when I make the claim that all you have to do to detect solar neutrons is detect several argon atoms in 100,000 gallons of cleaning fluid! It is great fun to then point out that this experiment has been done, and moreover that no science fiction writer in his or her right mind would introduce such a notion in a screenplay because it seems so implausible. And without the *Star Trek* hook, I am not sure I would even have had an audience in the lecture theater to have this kind of "aha" experience, as they say in science museums.

SCIENCE FICTION INTO THE CLASSROOM

But can this approach work beyond the world of popular books and lectures, and in the classroom? I believe it can. Since writing my books, I have heard from countless high school teachers that they have used *Star Trek* or other examples from science fiction for some time as a way to both motivate otherwise uninterested students and to further excite those students who are already turned on by physics. Having books that touch on the most modern developments in physics in this context helps teachers by giving them access to examples of which they would probably otherwise not have been aware.

Nevertheless, I have found that realizing that something very basic can be understood in a new way is often far more powerful than the satisfaction gained from obtaining new perspectives on the various exotica of modern physics. We all try to make the questions in our problem sets more exciting than the "Joe and Jane were traveling down the road at fifty miles per hour . . ." type that many of us were exposed to in introductory physics courses. Why not then get students to show that Jean-Luc Picard, captain of the *Enterprise* in *Star Trek: The Next Generation*, would be squashed like an ant by g-forces every time he uttered "Engage!" or that the alien invaders in the movie *Independence Day* would wreak havoc merely by bringing a spacecraft with a quarter of the mass of the Moon into a geostationary orbit around the Earth? It is for precisely these reasons, in fact, that I began my last two books with these two examples.

Indeed, there is a school of thought in physics education that suggests that the only way to really have students learn things and remember them afterwards is to make them directly confront their own incorrect preconceptions about physics. Get them, for example, to first explain why objects of different mass fall at different rates, and then show them that this isn't so. A generation reared on *Star Trek*—and more recently *The X Files* and movies such as *Star Wars* and *Independence Day*—is primed with misconceptions just waiting to be exploited!

DANGERS AND PITFALLS

This approach is, of course, not without its pitfalls. I have found that the most egregious public misconception about science is the feeling that scientific revolutions do away with all that came before them. The public thinks that nothing is impossible and that everything that we think is true today will one day be proved wrong. (So, the logic goes, why bother paying attention to physics at all?) This, needless to say, is completely antithetical to the central features of science—namely, that we can only prove things to be false (not true) and that principles that violate experimental tests now will continue to violate them in the future. (I wish I could convey these ideas more effectively to some of my postmodernist colleagues in the humanities!)

You have to be careful when having fun with the universes of science fiction to make sure that your audience does not come away confused about these central themes. Indeed, I receive more letters about ideas that I claim are impossible than anything else. (Most of the letters are prefaced by something like: "They would have said you were crazy if you talked about airplanes in the 16th century. . . .") Our responsibility is to teach people how to separate the difference between areas where we simply do not know the answers—where, indeed, anything or almost anything may be possible—from those areas where we have a clear notion of which ideas are incorrect.

While I recognize that resorting to the world of science fiction can sometimes blur this distinction, I think that, if one is careful, the advantages of reaching out to popular culture do, in the end, far outweigh the possible dangers.

Two Modest Proposals Concerning Scientific Literacy

James Trefil

I'm sure that everyone contributing to this volume has his or her own special memory of Leon Lederman, whether of his persona as a street-smart, wise-cracking New Yorker or of his (somewhat) more dignified incarnation as a national leader in science and education policy. For me, that special memory concerns a science school for federal judges where we were both lecturing. Over after-dinner drinks with some of the judges, the subject turned to the question of why someone like Lederman, Nobel laureate and director emeritus of one of the world's great research laboratories, would want to get involved in something as dirty and messy as the Chicago public school system. "Having a Nobel Prize will get you into anyone's office *once*," he explained. "I figured I could use that to do some good." This mix of sardonic humor and idealism is what we have come to expect from the man we are honoring with this volume.

There are many aspects of science education, each posing its own special problems. The problems in primary and secondary education that Lederman has chosen to tackle are discussed at length in other contributions in this volume. They have the advantage of being well defined and largely concerned with curriculum reform and teacher training; the disadvantage is that one is dealing with young students who often do not appreciate the value of what they are studying. The problem of scientific literacy, the area in which I have chosen to concentrate my efforts, is in some sense a mirror image of the primary and secondary education field. The good news is that one is dealing with older students and adults; the bad news is that they are either finished or at the very end of their education, and standard curriculum reform has limited effectiveness. To put it bluntly, people who have finished their formal education don't have to study for the final anymore.

In addition, progress toward scientific literacy through curriculum reform is always slow. Think of it this way: If we could wave a magic wand and suddenly have an educational system that performed flawlessly in science education, producing graduates with the highest levels of scientific literacy, it would still be a quarter-century before those graduates constituted a majority of the electorate. This is why so many of us (Leon included) spend so much time writing books, talking to people in the broadcast media, and worrying about the public face of science.

Nevertheless, the only long-term solution to the problem of scientific literacy involves changing the educational system. Because of this, I would like to spend the balance of this chapter talking about some aspects of science education at the university level. Before doing so, however, let me take a moment to define the term *scientific literacy,* as I will use it.

I will say that a person is scientifically literate if he or she can deal with scientific matters that come across the horizon of public life with the same ease an educated person would exhibit in dealing with matters political, legal, or economic. In a society that is becoming increasingly driven by science and technology, a society in which the citizenry is increasingly called upon to deal with issues that contain a large scientific or technological component, this kind of literacy isn't a luxury—it's a necessity. Without it, our democratic system would degenerate into one in which decisions are made either by an intellectual elite or by demagogue-driven mobs.

Having said this, however, it's important to realize that the sort of knowledge that a citizen needs to function in the twenty-first century is markedly different from that needed by someone planning a scientific or technological career. The average citizen doesn't have to be able to *do* science, just understand enough to come to conclusions about public issues. Take any of the current debates about biotechnology as an example. The questions being debated deal largely with moral and religious issues, but it's clear that someone who doesn't know what a gene is or understand something about the basic molecular mechanisms that drive living systems really can't understand what these questions are. This person is, effectively, cut out of the democratic debate. On the other hand, you don't have to be able to sequence a segment of DNA in order to decide whether (for example) therapeutic cloning does or doesn't fit with your moral calculus.

Thus, the concept of scientific literacy imposes a kind of "building code" on the educational system. It says that no one should be allowed to graduate from high school or university without a certain basic understanding of the workings of the universe. It imposes a standard, however, that is extraordinarily difficult for modern universities to deal with, since it requires an education that doesn't fit very well with departmental structures. The debate on global warming, for example, involves an understanding of planetary energy balances (physics), molecules in the atmosphere (chemistry), climate change (earth sciences), and the strengths and fragilities of ecosystems (biology). A student would have a hard time acquiring such a background in the types of courses that now dominate our curriculum—courses that concentrate on a single discipline.

I believe that universities can overcome their organizational inertia and produce the kind of multidisciplinary courses needed to produce scientifically literate students. Indeed, I know of at least 200 colleges and universities in the country that have either adopted or are experimenting with these kinds of courses. There is, however, another obstacle that I perceive to be much more fundamental and difficult—a problem that has to do with the attitude of many science faculty members. Too often, I find my colleagues encased in a kind of arrogance that makes it difficult for them to see any value in teaching students who are not destined to be future scientists or engineers. When these values are passed on to younger faculty and incorporated into promotion and tenure decisions, they produce a kind of collective blindness to the needs of ordinary citizens.

In the spirit in which Leon Lederman often initiates debate by making statements that are at the same time outrageous and eminently sensible, I offer two proposals to guide faculty efforts in the quest for scientific literacy. They are:

You have to teach the students you have, not the ones you wish you had.

If you expect students to know something, you have to tell them what it is.

Let me start with the first principle. No one who has spent time in faculty clubs can have failed to notice that one of the favorite pastimes among professors is complaining about the current generation of students. This is particularly true, I have noticed, among scientists who teach courses for nonscience majors. (I should mention in passing that the phenomenon of faculty griping seems to be independent of the actual quality of the students. I have listened to pretty much the same litany from people who teach at prestigious private institutions with extremely rigorous entrance standards and people who teach at community colleges. In fact, there may even be less griping from the latter than the former!)

Some of this complaining may be no more than the age-old tendency of elders to focus on the faults and failures of the next generation. Forgive me, however, if I see it as a symptom of something deeper, particularly as it relates to science education. It has been my observation that science faculty members, with some outstanding exceptions, are uncomfortable teaching their fields to those who do not intend to pursue the study of science at a higher level. In a sense this is only human—we all feel more comfortable with people who are "like us" (or at least who aspire to be "like us") than with those who are "not like us." In the case of science faculty, this unease takes the form of attacks on any course that treats a subject with less rigor than would normally be presented to majors studying the equivalent material. It might take the form, for example, of decrying the departmental offerings for liberal arts majors as watered-down versions of the real thing.

I feel that this sentiment is really an attack on the notion that any presentation of a science at less than full rigor is somehow inferior, and (by implication, at least) should not be in the curriculum of a real university. In this viewpoint, the fact that all students need to know *something* about science is irrelevant. "Either learn our subject in its most rigorous form," the argument goes, "or don't bother taking it at all." It is as if a music department decided you couldn't come to a concert unless you could prove your proficiency on the violin.

Obviously, addressing this subject in this volume is like preaching to the choir. Suffice it to say that the real obligation of a university faculty member (or any faculty member, for that matter) is to take the students that come into his or her class from wherever they are at the beginning to the place where they are supposed to be at the end. In the case of liberal arts students, that means that we have to accept the fact that we will be teaching people who have difficulty thinking quantitatively, who may not have a clear notion of what constitutes proof, or who simply have difficulty thinking in terms of physical models. Our goal should be to take these students—not the students we wish we had—and make them scientifically literate.

The second proposal concerns what happens after we have accepted this challenge, because it deals with the way we try to go about reaching our goal. There is a common divide among those engaged in scientific literacy, roughly characterizeable as a divide between advocates of *content* versus advocates of *method*. At the extreme ends of the spectrum, method people argue that the important thing is for the student to understand how science works, while at the extreme content end, the argument is that the most important thing is for the student to understand what scientists have discovered about the universe we live in. Obviously, the correct answer to the problem of scientific literacy lies some-where between these extremes, but everyone falls closer to one end of the spectrum than the other. (For the record, I am closer to the content end.)

In a sense, my second proposal is designed to counter what I see as an excessive move among modern educators toward the method end of the spectrum. In its most extreme form, this move works directly against the goal of scientific literacy. At the risk of setting up a straw man, let me state the method position as follows: There is something called the *scientific method*, and someone who understands this method will be able to understand all of science, regardless of the specific subject matter that person has been taught. Thus, the goal of science education should be to teach that method.

It's hard for me to understand how anyone could hold a position that is so clearly untenable. If I were to argue that there is a *linguistic method* that allowed one to learn languages so that it is irrelevant whether the student studied French or Macedonian, no one would take me seriously. I would argue that knowledge of one area of science gives no more insight into another than knowledge of French gives to Macedonian. I occasionally characterize the method approach, in fact, as the "teach them molecular biology and they'll work out general relativity on the bus home" school of thought.

The fact of the matter is that in order to be scientifically literate, the citizen needs to know a little bit about every field of science. He or she also needs to know a little about the scientific method as well, but that can't substitute for knowledge of the basic facts. A student in today's university, taking the standard eight-hour science area requirement, can easily graduate from college without ever having heard words like "DNA," "alternate energy source," or "food additive" in a classroom setting. This student, I suggest, has been poorly equipped to face the decisions that he or she will have to make as a citizen.

In short, if we expect students to know a little bit about every branch of science, then we have to *teach* them a little bit about every branch of science. There is no magic bullet that will allow us to keep teaching our comfortable departmental courses and still give our students what they need. We have no alternative but to start to transcend the nineteenth-century categories around which our disciplines are organized and become multidisciplinary in the best sense of that word.

Leon Lederman, in his work to bring together modern cosmology and particle physics, has already shown that this idea can be made to work in a research setting. Why shouldn't the next step be to carry it out in our teaching?

So ENTER THE "CIVIC SCIENTIST" WHO, LIKE SUPERMAN, MUST TAKE OFF LAB COAT, DROP HIS SLIDE RULE AND GET INVOLVED

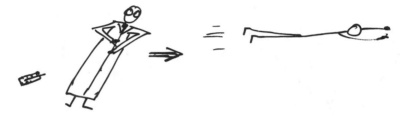

THE BATTLE IS FOR THE HEARTS AND MINDS OF THE CITIZENS OF OUR NATIONS. IT IS TO ENLIST THE POPULATIONS IN THIS COMPLEX SET OF ISSUES THAT CONNECT OUR SCIENTIFIC THINKING AND CAPABILITIES TO THE PHYSICAL, SOCIAL AND MORAL PROBLEMS THAT HUMAN CIVILIZATION FACES.

THE WAR IS A WAR AGAINST IGNORANCE. ITS BATTLE GROUNDS ARE THE SCHOOLS OF THE NATION

PART 4
SCIENTIFIC STEWARDSHIP

THE ETHICAL RESPONSIBILITIES OF SCIENTISTS

Howard Gardner

In the middle of the nineteenth century, a serious proposal was made to close the U.S. Patent Office because all inventions of significance had been made. In light of the subsequent appearance of the telegraph, telephone, radio, television, airplanes, and computers, we now laugh at the naïveté of this proposal. A few years ago, an American journalist named John Horgan wrote a book entitled *The End of Science* (1996). In this book, he speculated that the important questions about the nature of matter and life had been answered, and that most remaining questions about nature and mind were not susceptible to scientific answer. A century from now, the suggestion that science was at an end in the 1990s will seem equally ill informed.

To be sure, we cannot predict particular advances in science and technology. At the end of the nineteenth century, who could have anticipated such discoveries as the theory of relativity or plate tectonics; who could have anticipated quantum mechanics, the implications of Heisenberg's indeterminacy principle, and the work in particle physics carried out by Leon Lederman and others at CERN, Nevis

"The Ethical Responsibilities of Scientists," by Howard Gardner first appeared in *Dialogues*, January 1999. Used with permission, copyright © 1998 by H. Gardner.

Laboratories, Brookhaven National Laboratory, and Fermi National Accelerator Laboratory (Fermilab)? Turning from the physical to the biological world, who could have foreseen the revolution in molecular biology: the nature of genes and chromosomes and the structure of DNA, let alone the fact that we can now clone entire organisms, transform the human genetic sequence, and, if we wish, control our heredity? And now that significant progress is being made in the neural and cognitive sciences, it seems highly likely that we will continue to unravel the mysteries of thinking, problem-solving, attention, memory, and—the most elusive prize of all—the nature of consciousness.

For those who are close to science, it is hard to deny the excitement of the enterprise. So many issues and questions that were once the lot of poets and armchair philosophers have already been answered by scientists or are at least within their grasp. As it has sometimes been put, mysteries have now become problems, and problems are susceptible to solution. And yet, it is dangerous to adopt a pollyanish view of science. Science marches on. There is no guarantee that science will naturally contribute to the good of the public or that it will be a benevolent force in the future. As Leon Lederman (1992) once commented, "In the early days of science, as we look back on it, science had devastating effects on how people lived. By devastating I don't mean negative. I mean just dramatic changes in how people lived, but it wasn't known at the time that that would happen."[1]

Science is morally neutral. It represents the best efforts of human beings to provide reliable answers to questions about which we care: Who are we? How did we come to be? What is the world made out of? Where did it come from? What will happen to it? When? (Should I scribble the date on my calendar?) What determines the regularities and the irregularities in the world? What kind of creature would ask such questions? Is that creature moral, immoral, or amoral?

But what happens as these questions are answered? Sometimes, the answers simply satisfy human curiosity—a very important goal. But at other times they lead to concrete actions—some inspiring, some dreadful. Einstein's $E=mc^2$ (admittedly by a circuitous route) stimulated many outcomes. These ranged from the use of nuclear energy to power cities to the detonation of nuclear devices at the cost of thousands of lives in Hiroshima and Nagasaki to the spreading of fallout following the Chernobyl disasters. Following the discoveries of antibiotic agents, we behold the emergence of wonderful drugs that can combat dread diseases as well as the emergence of new toxic entities that prove immune to the effects of antibiotic medication.

Again, science itself cannot decide which uses to pursue, which not. These decisions are made by human beings, acting in whichever formal and informal capacities are available to them. Einstein is a good case in point. It is doubtful that he thought about applications of atomic theory when he was developing his ideas about the fundamental properties of matter and energy. When the politically attuned physicist Leo Szilard approached him in the late 1930s, it had already become apparent that nuclear energy could be harnessed to produce very powerful weapons. Einstein agreed to sign a letter to President Franklin Roosevelt and that action, in turn, led to the launching of the Manhattan Project and the building of the first atomic weapons. After the end of the Second World War, and following the detonation of nuclear devices over Japan, Einstein became a leader in the movement toward peace and eventual disarmament.

In the past scientists argued that their job is to add to permanent human knowledge and understanding, and not to make decisions about policy and action. But what forces, then, have prevented the random use, misuse, or frank abuse of technology—the fruits of scientific progress? What has been the role of scientific leaders—individuals like Leon Lederman, whose scientific achievements are a matter of history, but who have elected to address broader social issues?

We can identify three factors that have traditionally served as a restraint on the misapplications of science. First of all, there have been the values of the community, in particular, religious values. One could in principle conduct experiments in which prisoners are exposed to certain toxic agents. But religion counsels the sanctity of all human life. A second balancing force has been the law. In many nations, for example, prisoners are protected against unusual forms of treatment or punishment. Third, there is the sense of calling, or ethical standards, of professionals. A scientist can take the position that a contribution to knowledge should not be secured at the expense of human or animal welfare; indeed, some scientists have refused to make use of findings obtained by the Nazis as a result of immoral experiments. Or the warden of the prison may also refuse to allow his prisoners to participate in such studies, even if there are pressures to do so.

Each of these factors remains operative but, alas, each seems reduced in force nowadays. At a time of rapid change, values are fragile and religious values may seem anachronistic. Laws remain, unless they are overturned, but often events change so quickly that the law cannot keep up—witness the confusion of the United States Congress as it attempts to deal with issues like cloning and stem cell research. And during an era when the market model has triumphed in nearly every corner of society, it is often quite difficult for individual professionals to uphold the standards of their calling. A decade and a half ago physicians in France colluded in the sale of blood that they knew to be tainted by HIV virus; it is probable that their sense of calling was not potent enough to combat financial and societal demands for the blood.

We encounter an impasse. On the one hand, science and innovation proceed apace, ever conquering new frontiers. On the other hand, the traditional restraints against wanton experimentation or abuse appear to be tenuous. Must we leave events to chance, or is there a way to pursue science in a responsible way?

Enter the ethical responsibilities of the scientist. I contend that a new covenant must be formed between the scientist and the society. Society makes it possible for scientists to proceed with their work—by the funding of science and also by cooperation in its execution. In return, I submit, scientists must take on an additional task: They must relinquish the once justifiable claim that they have no responsibility for applications and undertake a good faith effort to make sure that the fruits of science are applied wisely and not foolishly. They may do so principally in two ways: (1) by focusing on the possible applications or misapplications of their specific research, and (2) by focusing on the relationship between the practice of science and the larger society in which it is situated.

Let me introduce an example from my own work as a cognitive psychologist. Nearly twenty years ago, I developed a new theory of intelligence called the theory of multiple intelligences (Gardner 1983). While I thought that this theory would be of interest primarily to other psychologists, I soon discovered that it was of considerable interest to educators all over the world. Educators began to make all kinds of applications of the theory. I was intrigued and flattered by this interest. Yet, like most scientists, I felt little personal involvement in these applications. Indeed, if asked, I would have responded, "I developed the ideas and I hope that they are correct. But I have no responsibility for how they are applied—these are 'memes' that have been released into the world and they must follow their own fate."

About ten years later, I received a message from a colleague in Australia. He said, "Your multiple intelligence ideas are being used in Australia and you won't like the way that they are being used." I asked him to send me the materials and he did so. My colleague was absolutely correct. The more that I read these materials, the less I liked them. The "smoking gun" was a sheet of paper on which each of the ethnic and racial groups in Australia was listed, together with an explicit list of the intelligences in which a particular group was putatively strong and an accompanying list of intelligences in which they were putatively weak.

This stereotyping represented a complete perversion of what I personally believed in. If I did not speak up, who would? Who should? And so, I went on television in Australia and criticized the program as *pseudo-science*. That critique, along with others, sufficed to result in the cancellation of the project.

I do not hold myself up as a moral exemplar. It was not difficult to appear on a television show in a faraway country, and I was not doing work in biotechnology or rocket science. Yet, the "move" that I made in my own thinking was crucial. Rather than seeing "applications" as the business of someone else, I had come to realize that I had a responsibility to make sure that my ideas were used as constructively as possible. And indeed, ever since that time, I have devoted some of my energies to supporting educational work on multiple intelligences of which I approve, and critiquing or distancing myself from work whose uses are illegitimate or difficult to justify. And to the extent that I am able, I have also begun to work on educational reform more broadly— indeed, it is in that context that I, a social scientist, first had the privilege of meeting Leon Lederman, a Nobel Prize–winning physicist.

How can one begin to forge a new covenant between the scientist and the larger society? To my mind, the current impasse calls for greater efforts by each party to make clear its needs and its expectations. Scientists must continually be willing to educate the public about the nature of science and what is needed for good scientific work to be done. Scientists have a right to resist foolish misunderstandings of their own enterprise and to fight for the uncensored pursuit of knowledge. At the same time, scientists must be willing to listen carefully to the reservations of nonscientists to their work, to anticipate possible misapplications of the work, and to speak out forcefully about where they stand with respect to such reservations, uses, and misapplications.

Ordinarily, neither scientists nor the general public should block the road of inquiry. Assuming that they do not harm others, scientists must have the right to follow questions and curiosity where they lead. Occasionally, however, scientists may want to consider not doing certain studies, even though they may be personally curious about the outcomes. In the case of my own field, I myself do not condone studies about racial differences in intelligence because I think that the results of these studies are likely to be incendiary. Some biological scientists are extremely reluctant to engage in experiments of genetic engineering or cloning of human beings, not because of lack of curiosity about the results, but rather because some of the implications of this work could be very troubling—leading, for example, to serious psychological or medical problems in the subjects of these experiments.

If one believes that my claim has merit—if one believes that scientists should become more deeply involved in ethical considerations—how might scientists act upon that belief? This is the question I have been pondering with my close colleagues Mihaly

Csikszentmihalyi of the University of Chicago and William Damon of Stanford University. We are trying to understand how leading practitioners—individuals doing "cutting edge work"—deal with the various invitations and pressures in their domain. We have been observing and interviewing scientists and professionals in other domains, such as journalism, theater, and philanthropy. We want to know how their present work situation appears to such individuals *in the trenches*; and we want to identify individuals and institutions that have succeeded in melding innovative work with a sense of responsibility for the implications and applications of that work (Gardner, Csikszentmihaly, and Damon 2001).

While it is too early to report the results of this work in any detail, I can mention a few tentative findings and the way in which we are currently conceptualizing the issue. To begin with, professionals are not naïve about their situation. They are aware of the great pressures on them and the hegemony of the market model at the turn of the millennium. They want to be ethical persons in their professional and private lives and they recognize that there are pressures that make it difficult for them always to do *the right thing* and to avoid crossing dangerous lines.

Yet clear differences can be observed in how successful these innovative individuals are in maintaining an ethical sense. Not surprisingly, early training and values are important, and that includes a religious affiliation in many senses. The opportunity to work in the laboratory of an ethical scientist, or to have other close colleagues with impressive values, is an equally important formative factor. We speak about vertical support—the opportunity to work with a senior *good worker*—and horizontal support—the opportunity to be surrounded by peers who also strive to carry out good work.

Once one has begun one's career in earnest, a creative individual is aided by two factors. The first is a strong sense of internal principles—lines that one will not cross, no matter what. If a scientist says—and believes—that he will never put his name on a paper unless he has reviewed all of the data himself, that virtually eliminates the likelihood that he will be an accessory to the reporting of fraudulent data. The second factor is a realization that the profession does not have to be accepted the way that it is today; as a human agent, one can work toward changing that domain. Suppose, for example, that it has become routine practice, in the writing of grants, for the head of a laboratory to propose work that has in fact already been carried out. A scientist can decide henceforth not to do so and work with colleagues to change the procedures in the domain. And indeed, the installation of a process where senior scholars apply for support by describing work that has been completed, rather than work that might be done in the future, would represent a significant alteration in the customary practices of a domain.

Similar examples can be gleaned with reference to the applications of one's work. One can decide, for example, that all of one's work is in the public domain and thus refuse to patent any findings. Here an internal principle gains out over the desire for personal profit. Or one can move toward the expansion of science so that it takes into account the public interest. One way to do it would be for every laboratory voluntarily to set up an advisory committee, consisting of knowledgeable individuals from other domains and laboratories. This advisory group would inform itself about the work of the lab, critique it when appropriate, and make suggestions about benevolent and possibly malevolent uses of findings.

Crucial in the pursuit of good work is the presence of individuals who embody good work in a given domain. We call such individuals *trustees* because they devote significant efforts to the preservation of the domain, not for personal gain, but rather in the best, most disinterested sense of that term. I cannot speak knowledgeably about Leon Lederman's contributions to particle physics but I can vouch for his generous gifts to the education of young Americans in the areas of science. During recent years, few if any top flight scientists have devoted as much attention to the ways in which Americans learn about science. Leon Lederman has done this in an institutional way—by reconceptualizing the science curriculum in secondary school and by playing an instrumental role in the founding of bellwether institutions, like the Illinois Mathematics and Science Academy (IMSA) in Aurora and the

Teachers Academy for Mathematics and Science (TAMS) in Chicago. And he has done it as well in a personal way, spending countless hours working directly with young people—those who are disadvantaged as well as those who are privileged—and introducing them firsthand to the excitement and the joys of active participation in the scientific enterprise. In these and other ways, Leon Lederman is a prototypical *trustee*, an exemplary *good worker*, who contributes directly to individuals and institutions and who inspires others to do so as well.

In the end, in my view, every individual has a set of four responsibilities. The first responsibility is to one's self—one's own goals, values, and needs—both selfish and selfless. The second responsibility is to those about one—one's family, friends, and daily colleagues. The third responsibility is to one's calling—the principles that regulate one's profession—in this case, what it means to be a research scientist. The fourth responsibility is to the wider world—to individuals one does not know, to the safety and sanctity of the planet, and to those who will inherit the world in the future. By dedicating himself to the education of young individuals, Leon Lederman embodies the famous sentence of Henry Adams: "A teacher affects eternity; he can never tell where his influence stops."

Whether sage or scientist, lawyer or layperson, all of us must negotiate our way among these strong and sometimes competing responsibilities; we are helped by religion, ethics, friends, and colleagues, but in the end we must do the balancing ourselves. Personal responsibility cannot be delegated to someone else. Those who have the special privilege of conducting science have a special obligation to be reflective about these competing responsibilities. And in a day when scientists have a strong handle on the nature of matter, sources of energy, the structure of life, and the means for creating and changing life, these responsibilities are awesome. Much greater mindfulness about this situation has become a necessity if we are to pass on to our progeny a world that is worth inhabiting.[2]

NOTES

1. The quotation is from an interview that Leon Lederman gave at the "Winding your Way through DNA" symposium, University of California at San Francisco, 1992. A partial transcript is available at http://www.accessexcellence.com/AB/CC/lederman.html.

2. Preparation of this paper was supported in part by the Christian Johnson Endeavor Foundation, the Louise and Claude Rosenberg Jr. Foundation, and J. Epstein Foundation.

REFERENCES

Gardner, H. *Frames of Mind: The Theory of Multiple Intelligences.* New York: Basic Books, 1983/1993.

Gardner, H., M. Csikszentmihaly, and W. Damon. *Good Work: When Excellence and Ethics Meet.* New York: Basic Books, 2001.

Horgan, J. *The End of Science.* Reading, Mass.: Addison-Wesley, 1996.

SCIENTIFIC RESPONSIBILITY

Walter E. Massey

Scientists in modern society have a variety of responsibilities stemming from their multiple roles as scientists, as public citizens in an increasingly global world, and as members of their local communities. The primary responsibilities of scientists are to maintain the health and vitality of science itself, to advance the discovery and dissemination of knowledge, and to ensure that the scientific enterprise continues to progress.

Other individuals and institutions also have important roles in this effort: government's support of research and education, the lay public's appreciation of science and scientists, and the media's efforts in publicizing and disseminating scientific news. But the core, indispensable responsibility for the health and progress of science rests, primarily, on scientists themselves. This may appear to be a narrow, self-serving goal. However, I will argue that in order to achieve this goal, scientists must engage in other activities that are beneficial to society as a whole.

Specifically, there are four important efforts and activities scientists should be engaged in and support so that the health and vitality of science is maintained.

First, the best and brightest minds must be attracted and encouraged to participate in science, which also means attracting people from all cultural and ethnic backgrounds.

Second, as a scientific community, it is important that we participate in efforts to enhance the quality of science education for all people, thus helping to establish a more scientifically literate populace.

Third, we must insist that the scientific community, in the conduct of research, teaching, and dissemination, adheres to the highest standards of integrity and ethical behavior.

Fourth, we need to ensure, as well as we can, that scientific knowledge is used for the benefit of humankind, not to its detriment.

USE KNOWLEDGE FOR HUMANKIND'S BENEFIT

This fourth point is, if not controversial, at least highly debated. But it has become more relevant today than perhaps ever before. Historically, there has been an ongoing debate as to the degree scientists can attempt to control the outputs of their research. I think, however, there are a diminishing number of scientists today who would adhere to the principle that science's only role is in discovery and advancement of knowledge and that "society has to determine the uses to which the results of science are put."

Current debates surrounding issues such as cloning, the use of stem cells for research, uses of various forms of energy, including nuclear energy, and the scientific basis for global climate change are all issues about which the public and policymakers expect the scientific community to help them in reaching responsible decisions. As difficult as it might be, the scientific community will have to assume an increasingly public responsibility in attempting to mitigate the harmful aspects of new discoveries and breakthroughs and to maximize the benefits to society. The public will expect no less.

SCIENTIFIC COMMUNITY NEEDS HIGH STANDARDS OF INTEGRITY AND ETHICAL BEHAVIOR

My third point, the need for the scientific community to be explicitly ethical, is not controversial, but, nevertheless, needs to be continually stressed within the community and to the lay public. This is an issue of ongoing concern because there are instances, perhaps few, where individuals have not adhered to the accepted standards of the scientific community. Issues of misconduct in science (including plagiarism), altered results of scientific experiments, and the appearances of conflicts of interest have received widespread publicity.

In an essay in *Science* magazine titled "Taking Responsibility" (Rotblat 2000), Joseph Rotblat discussed his experiences in the development of atomic/nuclear weapons. He recounted his growing concern over the uses to which these weapons would be put. His experiences were similar to those of many other physicists involved with the Manhattan Project. Rotblat said,

> I am increasingly concerned about the role of science and technology, both in day-to-day life and in the destiny of humankind. Whether directly through the development of new military capabilities or indirectly through the uneven distribution of the benefits of new technologies, the future of civilization and the very existence of the human species are in peril. Scientists must bear responsibility for this danger and must take steps towards its removal. *Ethical considerations must become part of the scientists' ethos.* (Rotblat 2000; emphasis mine)

The lay public holds scientists in very high regard. In biannual surveys conducted by the National Science Foundation (National Science Board 2000), scientists are consistently rated among the most admired group—far above politicians. The public's admiration of scientists and its willingness to support science is based not only on its belief that science will benefit society but also that scientists can be trusted to be honest in conducting and reporting research.

Research pronouncements by scientists with apparent conflicts of interest (conflicts such as support by companies in their
field of research or ownership in companies that tend to benefit
from their research) are increasingly greeted with skepticism, if
not worse, by the media, government officials, and the public. The
problem is greater now than ever before because of closer
industry-university relationships and the rapid commercialization
of science. Much of the rapid economic growth in the U.S.
economy over the past two decades has been driven by technological developments that have been enhanced by the ability of
scientists to be more involved in the commercialization of their
research, from incubators on campuses to start-up companies
spun off from university research to joint projects between universities, national laboratories, and industry. All have made the
United States a leader in moving its investments in research from
the laboratory to the marketplace. However, this new environment
requires the scientific community to be more openly vigilant than
ever in adhering to the strictest standards of ethical behavior, and
in ways that are apparent and convincing to the public.

Equally important, if not more so, the scientific enterprise
itself cannot function without a strong, ethical base. The conduct
of science and the advancement of knowledge are predicated on
the fundamental assumption that research is carried out and
reported honestly, with all its flaws and uncertainties, and that
there is a community-wide commitment to seek the truth as best
we can understand it, regardless of the consequences.

In a volume called *Science and Human Values* (Bronowski
1965), Jacob Bronowski wrote, "Truth is the drive at the center of
science. Science must have the habit of truth, not by dogma, but
by process"—that is, truth must be inescapably woven into the
very fabric of science. Adhering to this code and transmitting it to
students is one of the most important responsibilities of scientists.

SCIENTISTS NEED TO PARTICIPATE IN SCIENCE EDUCATION

My second point concerning the responsibility of scientists to help enhance the quality of science education and to thereby, hopefully, establish a more scientifically literate populace, also is not controversial but also bears continuing emphasis. A scientifically literate populace is good for the progress of science and for helping to ensure that the results of science are beneficial to society. Without quality science education that extends to all parts of the populace, it is very difficult to achieve even a moderate level of scientific literacy.

By scientific literacy, I mean having a basic knowledge of science principles and facts and an elementary understanding of the scientific process. The National Science Foundation's biannual surveys on the public understanding of science show striking similarities in this problem, from the United States to Europe to Japan. In all of these societies, with some minor differences, the general level of knowledge about fundamentals of science is woefully inadequate. Scientists have a responsibility to help address this problem because of its importance to society as a whole (as the world becomes increasingly permeated by science and technology, citizens of all nations will need to have some basic appreciation and understanding of technology) but also because addressing this issue is important to the health of the scientific enterprise.

Individuals and organized groups increasingly demand to participate in scientific and technical decisions on issues such as the appropriate forms of energy to be used, global climate change, the use of stem cells in research, and so forth. If these individuals and groups are scientifically and technically literate, their involvement can be healthy for society. If they are not, serious problems for society, and for science, can evolve. Science literacy is important.

A scientifically literate and informed populace can better appreciate the complexity of scientific and technical issues, are less likely to be victims of political demagoguery, and more likely to participate in the civic and political process where scientific and public policy issues intersect. Furthermore, the scientifically informed public, much more than the general public, believes that "the results of scientific research benefit more than harm mankind" (National Science Board 2000). And, finally, a scientifically literate populace is more likely to vote for programs and policies (funding, etc.) that support and advance science. This section of the populace supports the public funding of scientific research more than the uninformed public (National Science Board 2000).

What might we reasonably expect a scientifically literate populace to know and be able to do? My simple criteria would be to expect that such a populace is skeptical, questioning, and curious. The goal of scientific literacy would be at least to minimize the influence of demagogues and charlatans, to reduce "credulousness." As mathematician and philosopher W. K. Clifford wrote in *Science and Human Values* (Bronowski 1965), "The danger to society is not merely that it should believe wrong things, though that is great enough; but that it should become credulous"—that is, that it will believe almost anything.

In this regard, it is important that scientists assume a responsibility to attempt to interpret science to the public in language and terms the public can understand. (And here, I hope I might be excused for inserting a Leon Lederman–type joke. A scientist is asked in his lecture to the general public to please interpret his results in layman's terms, and he says, "I'm afraid I don't know any layman's terms.")

SCIENCE NEEDS TO ATTRACT A DIVERSE POPULATION

Finally, let me address my first point, which is that the health and vitality of science depends on attracting the very best minds of people of all backgrounds, cultures, and ethnic and racial groups. Science is inherently a multicultural endeavor. Scientific ability, interest, and genius are not confined to particular racial, ethnic, or cultural groups. Historical evidence shows that breakthroughs in science and valuable contributions have been made by individuals from an outstandingly wide variety of cultural and ethnic backgrounds. For example, over the past thirty years, Nobel Prizes in physics have been awarded to scientists whose heritage is French, Japanese, Danish, Russian, Pakistani, Indian, Italian, German, Swiss, Chinese, Dutch, British, and "American," which comprises practically all national origins. The same pattern exists in other scientific fields.

Furthermore, science, at least in the United States, is not attracting a sufficient number of graduate students from the United States. In order to maintain the level of scientific participation needed for the health and vitality of the scientific enterprise, we, at least in the United States, have to reach out to bring into the field individuals from previously underrepresented groups in our country.

Perhaps an even more important reason for inclusion is that science is enriched by having individuals from different cultural and ethnic backgrounds because they bring to the scientific enterprise different viewpoints. They also may view physical phenomena through different lenses, and thereby enrich and broaden the way in which we interpret and understand the physical universe.

Professor Richard Nisbett at the University of Michigan provides some evidence that people from different cultures do see, understand, and interpret physical phenomena differently. Professor Nisbett and his colleagues have conducted extensive research on how individuals from different cultural and ethnic heritages develop and use *cognitive processes*—how they think about problems, sort evidence, and reach conclusions. Nisbett presents a large body of evidence that "literally different cognitive processes are invoked by different groups [when] dealing with the same problem" (Richard Nisbett, personal communication, manuscript in preparation).

Professor Gerald Holton, physicist and science historian at Harvard University, does not subscribe to the view that people from differing cultures view the universe "through different lenses," but he does strongly support aggressive inclusion of individuals from all groups in the scientific enterprise. In private correspondence to me, he wrote, "[It is] almost inevitable that we might capture novel or unusual insight into the understanding of the universe from people who have different life experiences and come from different cultures, simply because the larger the pool of well-trained and hardworking people, the larger the probability of novel and unusual insights. In this sense, excluding potential scientists *is a crime against the ethos of science itself"* (Gerald Holton, personal communication; emphasis mine).

Also, the inclusion of people from varied and different backgrounds may mitigate against the detrimental uses of science. The detrimental consequences of the application of scientific research are practically always unintended and unanticipated, at least by the scientific community (weapon research excluded). These applications are often the result of insensitivity, ignorance, or callousness toward the populations affected. One can make the argument that such detrimental consequences might be mitigated if scientists from the cultures, groups, and regions affected by the research were part of the research and application efforts. At least it seems reasonable that the risks of unanticipated consequences would be reduced by such broader inclusion.

In summary, then, of the many responsibilities of scientists, the most critical is to preserve and enhance the health and vitality of the scientific enterprise. The achievement of this goal will depend on the degree to which scientists do at least four things: (1) attract into science the best and brightest individuals from all cultures, racial and ethnic groups, (2) contribute to improving the quality of science education and enhancing science literacy, (3) insist on explicit ethical behavior in the conduct of science, and (4) to the degree possible, ensure that the results of science benefit rather than harm humankind.[1]

Happy Birthday, Leon!

NOTE

1. This essay is based on a paper given at the seminar on Science, Technology, and Society for the Twenty-first Century, Santander, Spain, September 2000.

REFERENCES

Bronowski, Jacob. *Science and Human Values*. Rev. ed. New York: Harper and Row Publishers, 1965.
National Science Board. *Science and Engineering Indicators*. Washington, D.C.: U.S. Government Printing Office, 2000.
Rotblat, Joseph. "Taking Responsibility." *Science* 289 (2000): 729.

IN 1600, FRANCIS BACON WROTE ABOUT
THE IMPORTANCE OF ORDINARY PEOPLE
TO UNDERSTAND THE PROGRESS OF
SCIENCE.   IN 1600,"IT WOULD BE USEFUL",
HE SAID.   IN 2000, IN A CENTURY
ALREADY EXPLODING WITH SCIENCE-BASED
TECHNOLOGY, I BELIEVE IT IS ABSOLUTELY
ESSENTIAL FOR SURVIVAL OF DEMOCRATIC
SOCIETY!

PART 5
BEYOND SCHOOLS
Demystifying Science for Public Policy

SCIENCE LITERACY AND SOCIETY'S CHOICES

Mae C. Jemison

*"Research is formalized curiosity.
It is poking and prying with a
purpose."*

Zora Neale Hurston,
in *Dust Tracks on a Road*, 1942

The United States is the most prosperous nation in the history of the world, and we are currently living in a particularly prosperous time. Although all societies should always be aware of the historical legacy that they are creating, moments such as this, when we as a nation are extraordinarily material-rich and relatively free from war and economic need, are especially propitious for performing deeds of lasting importance to the human race.

History's assessment of past societies is colored by their achievements in science and engineering and how they have improved the state of humanity. Societies that did not add to humankind's fundamental understanding of nature are spoken of less often. The past achievements of the United States in basic scientific and engineering research are major sources of national pride, comparable to our advances in political freedom,

"Science Literacy and Society's Choices," by Mae Jemison, is derived from a white paper from S.E.E.ing the Future, published by the Jemison Institute at Dartmouth College.

industrial and military might, and material well-being. Yet, the disturbing trend of the past decade has been to set aside future returns for short-term gains.

And nowhere is that trend to look for short-term answers and gains more dangerous than in the education of society—especially our long-term failure to ensure science literacy of our children, the general public, and our leaders.

THE IMPORTANCE OF SCIENCE AND ENGINEERING RESEARCH

Throughout history, human progress has been shaped, advanced, and thwarted by our understanding of the world around us and the application of that knowledge to create tools to achieve society's goals, whether stated or unstated. From our first success of reliably making fire to establishing formalized and standardized systems to barter and trade goods and services (money, taxes, financial institutions); from our concepts that illnesses affect the flow of energy in the body and application of poultices to wounds to the use of magnetic resonance imaging to study the *thinking* brain; and from honing rocks into spear points to the positioning of single molecules to create engines, again and again it is our knowledge, our understanding of the world around us and the application of that understanding, which has so significantly defined human society and humans' impact on our world.

Over the last century, tens of thousands of years into the course of human history, our ability to positively and adversely impact our very existence as a species on this planet has increased dramatically. And in just the last forty years our acquisition of knowledge and its subsequent application to build tools has accelerated at a startling, giddy rate. And so has our ability to negatively or beneficially affect the lives of everyone on this planet.

But exactly what is science and engineering progress? Science and scientific progress are not easy concepts to define. Journalists, the general public, and even many scientists often overlook the fact that the term *scientific* has meant different things at different times and in different contexts. How science is defined is always partly a social process. In turn, understanding the process of scientific progress has a critical influence not just on scientists but on society as a whole.

Francis Bacon defined scientific progress as banishing the accumulated errors of the past, the *idols of the mind*.

Scientific research attempts to help us understand the universe around us, the impact of our interactions with it. Modern scientific methodology generally requires that information added to our accepted knowledge base be independently observable by more than one person, and given the same set of circumstances is consistently reproducible. From analyzing these observations, scientists then go on to develop descriptions (hypotheses) of the world that can predict the outcome of an event (effect) and how it happens (cause). So progress occurs in fits and starts as new observations are made, new tools to test theories are developed, and new insights are gained.

Today, when understanding science is crucial to thousands of decisions at every level of society, relatively few Americans fully grasp the idea of scientific method. It is common to confuse the rigorous testing of hypotheses—part of the best tradition of modern science—with scientific uncertainty, or to think that a theory remains *unproved* as long as some scientists can be found to disagree with it.

In political or economic debates, in considering environmental or energy policy or missile defense systems, both sides typically enlist scientists to support their arguments. Can Americans filter out the science from political bias and financial self-interest?

Our future as a nation and society may depend—more than we realize—on our understanding of the scientific method as a flexible, nuanced, and continually evolving path.

Much of the knowledge we have gained over the last half-century in the United States has been obtained through the use of public funds allocated to projects, universities, researchers, companies, and individuals by the government. Public funding of research in basic science and engineering in the United States led to the conceptualization, design, and initial implementation of the Internet; the development and launch of communication and weather satellites; and through biotechnology, the ready availability of human insulin to treat diabetes and erythropoetin to stimulate red blood cell growth in cases of severe anemia. These benefits, though just recently available, are the direct and indirect result of basic research funded by the public twenty to thirty years ago.

The public coffers that made such research possible—whether at the federal, state, city, or local level as well as profits from industry are derived from monies collected not only from corporations and businesses but from men and women who work daily at all the tasks that keep society running—high school teachers, lawyers, physicians, architects, nurses' aides, taxi drivers, airplane pilots, coal miners, baseball players, English professors, garbage collectors, musicians, cosmeticians, dish washers, and traffic cops, to name a few.

These members of our society as a whole may not always recognize the vital role basic science and engineering research play in their lives. Those allocating public funds for research, however, as well as those receiving them have the crucial responsibility of ethical stewardship on society's behalf. That is, as much as possible, the leaders in science and engineering research and policy must ensure that these public funds work to build a foundation that helps society reach its potential and supports society's ability to choose the best path.

"Research proceeds by making choices," Donald Stokes wrote.[1] Researchers themselves do not make all—or even most—of these decisions.

Research proceeds only with the support of larger society. If a scientist has the freedom to pursue his or her own interests and passions, it is only because a host of agencies and bodies—university departments, tenure committees, congressional legislation, corporate executives, market forces, government bureaucracies, defense strategists, city councils, school boards, state planning commissions—have tacitly endorsed those interests and provided the means to continue. Often without even realizing it, these forces push, prod, and cajole scientific and engineering research in one direction or another.

Those of us who consider ourselves researchers know all this intuitively, but even we don't always think through the implications or the questions. Will Adam Smith's *blind hand* guide genetic engineering as efficiently as it supposedly guides the larger economy? Are the *people's representatives* equipped to evaluate the importance of physics' string theory? Can local school boards dictate the facts of modern biology? Are democratic principles compatible with the free, unencumbered inquiry into the nature of the universe? How does scientific inquiry affect technological progress, and in what ways does technological progress benefit society?

Finally, can questions in the interest of society be answered, or even posed, before science and technology research has moved on to a new place beyond them? Only if our society ensures widespread and universal science literacy.

EDUCATIONAL GAPS AND OBSTACLES

In this century, universal science literacy is a requirement for a truly participatory democracy. Science literacy does not imply or require the ability to solve linear equations, recount the structure of DNA nucleotides, or elucidate theories on combustion. Rather, it is a baseline level of knowledge and skills that will allow a high school graduate to read a daily newspaper with information about health care, the environment, and new computers and to understand what it means for themselves, their families, and their community. Not an unreasonable expectation in 2002, but one that remains unfulfilled in the United States today.

What is the current landscape and what are the obstacles to achieving science literacy?

U.S. elementary and secondary school students score consistently below the international average for industrialized countries. U.S. student enrollment in higher education science and engineering classes and programs is declining. U.S. media coverage of science- and technology-related phenomena and advances is overwhelmingly weighted to disasters and pseudoscience and epiphenomena (crop circles, UFOs, haunting, and so forth) and rarely deals with basic science information in any detail.

America's public education system shows weakness in science and mathematics at the secondary school level. The gap in scientific and mathematical knowledge between U.S. students and their peers in competing economies has been well publicized, but it is worth repeating. Although statistics show some improvement since the 1970s, and younger students score above international standards, U.S. students in the final year of secondary school score well below the international average on assessments of general science and mathematics. More worrisome in the mid-1990s, U.S. twelfth-grade advanced science students performed below fourteen of sixteen countries in standard physics assessment and below eleven of sixteen countries in advanced mathematics assessment.[2] Studies continue to show that many U.S. students are taught mathematics and science by teachers who do not have degrees in those subjects. And many elementary school teachers have never taken a college-level science class.

This data suggests that by the time American young people have graduated from high school, the vast majority of them have already been lost to science and engineering careers and most are not science literate. On the college level as well, the idea of a firm grounding in science as essential to a well-rounded education has declined since the 1970s. Increasingly, even introductory undergraduate level science courses are the realm of science majors alone.

U.S. student enrollment in science and engineering programs is in decline. The social groups that historically provided talent for American science and engineering no longer fill the demand. Post–World War II American science and engineering was a club whose membership consisted of native-born, white, middle-class males and a smaller number of male West European immigrants. Although student enrollment in science and engineering doctoral programs at American universities has increased in recent decades, the majority of the increase can be contributed to foreign student enrollment. Of the foreign students who receive a Ph.D. in science and engineering, only 53 percent of these are employed in the United States five years after graduation.[3]

Opportunities for women and minorities in American science and engineering have grown enormously since the 1970s. The

United States is among the leading countries in the world in the proportion of science and engineering degrees earned by women (by 1996, women earned 46 percent of the degrees in the mathematical sciences and 47 percent of the degrees in the natural sciences).[4] Yet the structure of American science and engineering has only just begun to acknowledge these changes.

As one commentator concluded, as women and people of color enter the professions of science and engineering, change will be the order of the day. "Change will have to happen simultaneously in many areas, including the conceptions of knowledge and research priorities, domestic relations, attitudes in preschools and schools, structures at universities, practices in classrooms, the relationship between home life and the professions, and the relationship between our culture and others."[5]

The problem of higher education must be addressed with elementary and secondary education. The process of education begins before formal, professional institutional science training begins. Changes in the fundamental methods of education are needed. But nationwide basic improvements in the quality of teaching and school curricula in science and mathematics must be implemented.

Workforce gaps threaten U.S. leadership in high technology fields and industries. That workforce consists not only of engineers and scientists but of technicians, assembly-line personnel, and product developers. Decreasing numbers in enrollment in science and engineering programs and failure to educate a science-literate citizenry cannot support an ever-increasingly knowledge-based economy in the United States.

The state of U.S. science and engineering education has an enormous impact on a nation's competitiveness, economic growth, and overall vitality. To ignore the implications of current national trends in science and engineering education is to ignore an impending peril to the nation as a whole.

A less visible "science education gap" is also apparent in the United States—what might be called "science ignorance in everyday life." Despite the fact that Americans are proud of their scientific and technological history and progress, despite the increasing emphasis on technology and ongoing technological re-education, Americans seem largely unaware of the science that underlies their community, professional, and home life, not to mention those science impacts on national politics and economic activity. The American public's awareness of science and engineering is in direct conflict with the importance science and engineering has in American lives and lifestyles.

National polls suggest that half or less of the American public understands such key scientific facts and concepts as electrons are smaller than atoms, antibiotics do not kill viruses, lasers do not work by focusing sound waves, and it takes the Earth one year to go around the Sun—even though these concepts underlie the activities of normal, everyday life in our society. As recently as the late 1990s, only 16 percent of Americans were able to define the Internet.[6]

Even a cursory examination of the mass media shows that sensationalism and pseudoscience overwhelm serious attention to scientific research and debate on critical issues related to science. American newspapers run daily columns on astrology, but even national *papers of record* cover advances in astronomy mostly in the back pages of weekly supplements.

Most Americans get both general and science news from television. Yet much of the *scientific* coverage on American television networks is actually devoted to haunted houses, extrasensory perception, crime, alien abduction, crop circles, and unidentified flying objects. Although legitimate scientists sometimes appear in such programs, even so-called science networks rarely submit such stories to rigorous scientific reasoning—or, except in the case of forensic science, even to hard journalistic standards.

The Pew Research Center for the People and the Press tracked the "most closely followed news stories in the United States" with "at least some relevance to science and medicine." The top ten stories tracked since the 1980s included nine involving natural and man-made disasters (the *Challenger* explosion, Hurricane Andrew, Chernobyl disaster, California earthquakes, and similar events). Ten of the top fifteen studies were concerned with the effects of weather.[7]

Arguably more important topics ranked much lower on Pew's list of the top thirty-nine science stories: "the debate over U.S.

policy concerning global warming" (35), "discovery of scientific evidence of the beginnings of the universe" (36), and "the cloning of mice by scientists in Hawaii" (39).

Missing entirely from Pew's list of 689 closely followed stories were any involving advances in computer science (including the Internet), the impact of science on technology and the U.S. economy, human evolution, scientific study of the effects of drugs, endangered species, mathematics, genetically modified crops, or the study of human behavior.[8]

Although Americans have a high respect for science and four out of five agreed that encouraging the brightest young people to go into scientific careers should be a top national priority,[9] the popular image of science, scientists, and those who work with technology paints a very different picture. Scientific careers are perceived as difficult, obscure, and financially unrewarding. Scientific research is made to seem the preserve of a tiny group of rather peculiar and somewhat antisocial people. True understanding of science is deemed irrelevant to everyday life while the study of such epiphenomena as unidentified flying objects, psychic visions, and poltergeists are promoted relentlessly by the national, local, and regional media.

Such scientific and technological ignorance among the electorate surely translates into lack of concern by their elected representatives in Washington—and these are some of American science's most important patrons. That, in turn, identifies dangers ahead for America's scientific and technologic future, but also suggests opportunities for reeducation and change.

A SCIENCE LITERACY CAMPAIGN

What can be done? A federal science literacy campaign must be developed. This science literacy campaign should not just be directed at improving science education in schools, but also public understanding and awareness of the importance and process of scientific and technological support. The outreach should reach all segments of U.S. society and make us all aware of our collective responsibility for ensuring that human progress benefits from government investment in science and engineering research. This campaign, analogous to the War on Drugs, arguably will yield greater benefits and is as or more critical for achieving national and global security, real democracy, and human advancement than our current War on Terrorism, sparked by the attacks on the World Trade Center and the Pentagon.

Public funding of science and engineering research can have a tremendous influence on science and engineering education at all levels—formal elementary, secondary, university, and postgraduate programs and informal education to the general public. So in addition to funding science education initiatives from traditional sources—school boards and the Department of Education—the funding of science and engineering research represents significant potential for science education support. In fact, any agency considering public funding of science research should work from a philosophical foundation that promotes a positive, overall, long-term impact on education.

Effective science education must build students' interest and curiosity in science, engineering, and technology fields and foster the ability to digest and use information, not just demonstrate their ability to reiterate facts. Emphasis must be given to K–12 education. It is during the elementary grades that students begin to develop the basic skills and grounding that will allow them to become the technicians, engineers, and scientists of tomorrow. Elementary and secondary school is also when the lay public has its greatest and most important educational exposure to science. Projects that include hands-on, experiential, discovery-based approaches for students suitable to the targeted age should be given special consideration.

Programs should help inspire and channel children's interest in science careers, present diverse images of scientists, and take care to avoid stereotyping science and scientific careers. They should establish positive, real role models in a wide variety of science disciplines for K–12 education. This is especially important for minority, female, and underprivileged children. A program's design must take into account the need to measure its effect on education. Outcomes to be measured should include not only the numbers and types of student impacted but also the information gained, whether the program stimulated their interests in the science and technical fields and whether their critical thinking and problem-solving skills are enhanced.

Long-term partnerships between government funding agencies, industry, and K–12 education must be explored and encouraged. Industry can provide real-life practical experience for K–12 education and at the same time supplement needed resources in equipment, skilled workers, and funding that enhance K–12 science and engineering education. Many corporations already have significant programs in these areas that can serve as models and possibilities for such partnerships. Programs such as Bayer Corporation's Making Science Make Sense, Lucent Technologies' Project GRAD, Dow Chemical Company's Scientists in the Classroom, Intel's sponsorship of the International Science Competition and Talent Search, General Electric's ELFUN, NASA's student internships, and DOE's summer programs are some examples. These and many others demonstrate how projects involving professional scientists can partner with schools in hands-on science education.

An important side effect of engaging in research is the desire for more knowledge. Humans have a natural thirst for learning about the universe. Children, in particular, have it in abundance, but may not have beneficial outlets for it. Involved in their own lives and jobs, adults may feel that dealing with the significant scientific interests of their children are beyond them and are best left to specialists. Effective programs in K–12 should provide students with opportunities and outlets to take advantage of their curiosity and energy to explore the world around them. For example, field investigations with practicing engineers or collaborative projects that gather information for national databases, science fairs, and meeting with scientists provide such experiences.

Public funding must facilitate strong science education in geographic areas and demographic groups where it is lacking and hence most needed. Economically underdeveloped areas that have few local resources to augment their educational needs and opportunities should be major targets for funding of science. At the same time, national educational initiatives should also foster a trained workforce in geographic regions where there is a present shortage of skilled workers and where future needs cannot be met. Despite the importance of science and engineering research and science and engineering training to the whole of American society, the resources for research and training are by no means evenly distributed geographically. In fact, the six states with the highest levels of R&D expenditures (California, Michigan, New York, New Jersey, Massachusetts, and Texas) account for approximately half of the nation's total R&D outlay.[10] These areas typically experience severe shortages of technically skilled labor in good economic times while other regions of the country struggle to create and attract the high-paying jobs in technologically advanced industries.

Capturing, developing, and retaining the best minds from the United States to work in science and engineering fields is an important consideration of research funding. National initiatives should foster the recruitment and retention of all students, and emphasize traditionally underrepresented groups in the sciences. The diversity of ideas that result from the participation of these individuals maximizes the chance of breakthrough research, broadens public support, and increases the number of people ultimately entering the fields. Achieving diversity in science is a necessity, not a *nicety*.

The reason for retention or lack of retention of underrepresented groups at each level of education and training must be understood. Children across the spectrum of the U.S. population are excited by science and engineering, and adults understand its importance. But, sequentially at each educational level, groups traditionally not included drop out in higher and higher numbers. For example, of the undergraduate women who reach the threshold for the engineering path (completion of three required engineering courses) only 42 percent completed their degree as compared to 62 percent of men.[11] This lack of retention is a problem as demographics of the nation change and the pool from which the United States has traditionally drawn its science and engineering workforce shrinks to a smaller percentage of the labor force.

Senior faculty members, who are the major beneficiaries of funded research, should share the major burden and responsibility for improving science education outreach. Many funding agencies and programs already require investigators to submit a plan to increase public awareness and participation of underrepresented groups. The current system, however, seems to place this burden on younger faculty members and others just beginning their research careers. Public funding should thus find a way to encourage, in fact require, more senior, established faculty members to shoulder a significant portion of this outreach. Not only do they have the time and the job security, but frequently senior faculty has the additional overhead funds and institutional wherewithal to make such programs a reality.

The physical plants for scientific facilities must be effective, up-to-date, and capable of supporting the training and education of the myriad of personnel involved in science and engineering research and development. Schools in the United States that train the vast majority of our science and engineering technicians, professional engineers, and science workers are doing so on inadequate and often obsolete equipment. As they enter the workforce, the effectiveness of these workers is compromised until they receive compensatory training from the employer. Access to up-to-date laboratories and scientific equipment are essential not only to the research scientists themselves, but also the various lab technicians, administrators, and public servants who will assess, fund, and support research projects.

During the Cold War there was a perceived need and a concerted effort to improve science and engineering facilities across the United States. Over the last fifteen to twenty years, however, although federal funds for research at academic institutions have grown and diversified, there has been a distinct decline in science and engineering facilities at federally supported American colleges and universities.

For example, although the total academic science and research space increased by almost 28 percent between 1988 and 1998, the R&D equipment intensity—that is, the percentage of total R&D expenditures from current funds devoted to research equipment—

has declined dramatically in the past decade. After reaching a high of 7 percent in 1986, it declined to 5 percent in 1997.[12] While American colleges and universities are able to accommodate more science and engineering students, they are educating them with increasingly inadequate and obsolete equipment.

Except in the large, well-endowed research universities with access to large amounts of private funding, academic science and engineering facilities have also tended to be improved only to meet the needs of specific, funded projects. Additionally, schools that are not major recipients of government science and engineering funding typically receive less funding from other private sources to build science buildings and acquire research equipment. This trend has helped lead to a general deterioration and senescence of science and engineering laboratory facilities nationwide. Serious consideration must be given to creating substantial public funding initiatives that promote and ensure the vitality and currency of science and engineering laboratories in institutions throughout the country—at community and small undergraduate colleges and regional state university branches, as well as the major research universities.

In order to build and maintain a stable base of not only principle research scientists but also laboratory technicians, undergraduates at all colleges and universities should have some opportunity to do research. This does not imply that all institutions must be research universities; rather there should be outreach in some form to all schools so that their students are able to participate in research projects. Undergraduates majoring in science and engineering must have some exposure to research and access to research opportunities regardless of their school.

Current funding assessment protocols often favor factors found at colleges and universities that are already well established as centers of research and science and engineering education. But funding assessments should also consider the benefits of projects that target the incorporation of specific groups with less access to up-to-date facilities and training, and colleges and universities in the process of developing better facilities and a more qualified teaching faculty. Fellowships targeted toward undergraduates should continue to be used to encourage a diverse composition of the principal investigator's research team.

One of the main tasks of public monies in the United States is the education of children from grades K–12. A general appreciation of science is not only the starting point for helping young people aspire to be the next generation's scientists and engineers, it is also an essential ingredient in preparing America's citizens for life in an increasingly scientific and technologically advanced society. Basic science literacy, a broad societal understanding of and support for the role of science and engineering research, is essential. For a democratic society to function effectively, the citizens must be well informed and well educated about science, scientific research, and their implications.

Citizens are asked daily to consider individual and societal choices and policies that require basic science literacy and some familiarity with current scientific and technical discoveries and advances. Participation in or exposure to science and engineering research—from K–12 to higher education—will go a long way in helping the general public become more comfortable with the process of scientific research as well as give researchers a better feel for what the public needs and wants to know.

Formal education of American scientists and engineers, the education of the American workforce for the so-called knowledge economy, and public awareness of science are essential parts of the nation's scientific infrastructure. For every publicly funded research project, consideration should be given to a project's impact on training new scientists, technicians, engineers, and researchers, to educating a science-literate workforce, and to increasing public awareness of promoting science.

What does this all mean?

SOCIETY'S CHOICES

Right now, in 2002, the United States as a nation has the wherewithal to make profound, lasting, beneficial contributions to our citizens and all humanity through our science and engineering capabilities. To ensure that impact, the country must make conscious, informed choices as to how we invest our public resources in science and engineering. I believe one of the best investments suited to public funding in basic science and engineering research will be to actively pursue a nationwide campaign to improve science education with as much visibility as our wars on drugs, terrorism, or cancer. Because, of course, if we win the war for science literacy we will have made tremendous advances in all the others.[13]

NOTES

1. Donald Stokes, *Pasteur's Quadrant: Basic Science and Technological Innovation* (Washington, D.C.: Brookings Institution Press, 1997), p. 6.

2. National Science Board, *Science and Engineering Indicators—2000* (Arlington, Va.: National Science Foundation, 2000), p. 5-3.

3. National Science Foundation, *Science and Technology Pocket Data Book—2000* (Arlington, Va.: National Science Foundation, 2000).

4. *Science and Engineering Indicators—2000*, p. 4-4.

5. Londa Schiebinger, *Has Feminism Changed Science?* (Cambridge: Harvard University Press, 1999), p. 195.

6. *Science and Engineering Indicators—2000*, p. 8-11.

7. Ibid., p. 8-8. This study was done prior to the intense coverage of anthrax after the September 11, 2001, World Trade Center and Pentagon tragedies.

8. Ibid.

9. Ibid., p. 8-13.

10. Ibid., pp. 2-28 and 2-29.

11. Women, African-American, Hispanics, Native Americans, Pacific Islanders, and individuals from lower economic groups and rural geographic areas are underrepresented in science and engineering fields as compared to their percentage of the population at large or of the U.S. labor force. This disparity is greater in certain fields such as mathematics, physics, and engineering and as one climbs the ladder into graduate school and postgraduate education and fellowships. Women in 1997 were 23 percent of the science and engineering workforce but 46 percent

of the U.S. labor force. Blacks, Hispanics, and Native Americans who in 1997 made up 24 percent of the U.S. population, represented only 7 percent of the total science and engineering labor force. Ibid., chap. 3, "Science and Engineering Workforce" and chap. 4, "Higher Education in Science and Engineering."

12. See ibid., p. 6-2.

13. This essay derives from the White Paper generated by the S.E.E.ing the Future (Science, Engineering and Education) Workshop held at Dartmouth College, November 10–15, 2001. The workshop that I had the good fortune to convene and chair brought together over twenty-four leading thinkers in the sciences, engineering, and the arts—academia, industry and small business, writers, theologians, finance, physics, biology, and mathematics—to discuss the best uses for public money for funding basic science and engineering research. Time and time again this distinguished grassroots group of individuals returned to the necessity of science literacy for the general public, for the science and technology labor force and our leaders, and the stumbling blocks in our way.

SOME CURRENT CONCERNS

Missile Defense, the Future of Nuclear Power, and the Hazard of Loose Nuclear-Weapon Material in Russia

Richard L. Garwin and Georges Charpak

V ice President Dick Cheney's Presidential Development Group on National Energy Policy, reporting in May 2001, put nuclear power at the center of the United States' twenty-first century energy policy. We agree that nuclear power can supply an increasing percentage of U.S. energy needs. The Cheney report[1] argues the environmental and economic advantages of nuclear power as a sustainable energy source and calls for increased international cooperation and redoubled scientific research into the reduction and treatment of high-level nuclear waste.

We welcome the report's consideration of nuclear energy's problems and future potential and view it as an important vote of confidence. In past decades, the public's view of nuclear energy has been dominated by two particular events: the meltdown of a reactor core at Three Mile Island, Pennsylvania, in 1979 and a catastrophe in Chernobyl, Ukraine, in 1986, in which radioactive material from a power plant was ejected into the atmosphere and spread over vast regions of Europe and Asia. Although the specter of large-scale disaster and the unresolved question of the ultimate disposal of radioactive waste still loom large in American public opinion, nuclear plants currently provide 80 percent of France's electrical energy.

In France, electrical power in general, and the nuclear power industry in particular, is a monopoly of the state. It is well run and less subject to criticism in the French political system. Recent official reports, however, have for the first time at that level raised questions about the economic and environmental wisdom of certain choices, such as the reprocessing of spent nuclear fuel. The French nuclear industry never speaks of spent fuel as waste but as something to be put into underground laboratories. Yet not a single ton of spent commercial reactor fuel there (or in the United States) has been subject to ultimate disposition.

Considered against the fossil fuels—coal, oil, and natural gas—nuclear power offers the promise of cleaner air, reduced greenhouse emissions, sustainability, and the potential for increased output to match our escalating energy demands.

We strongly support energy conservation, with wider use of combined heat and power systems and hybrid (gasoline engine and battery) vehicles. We note that continued application of science and technology to solar energy will bring cost reductions. But solar and wind power are intermittent and require storage. They are extremely important for dispersed populations, but useful for large agglomerations only with transmission lines like those used for nuclear power. We have written a book about nuclear power and nuclear weapons; we discuss other energy sources simply to provide a context. We do quote authorities in the field to the effect that the conceivable solar-generated power worldwide could provide thirty times the world's current consumption of electricity. "More power to them!" we say.

However, to support a resurgence of nuclear power in the United States, a political endorsement must be matched by good science and favorable economics. We judge the 100 existing U.S. nuclear plants, all light-water reactors, to be adequately safe if operated properly. The disposal of spent reactor fuel could be handled by competitive, commercial, mined geologic repositories in Russia, China, the United States, and Australia, thereby separating the radioactive material from the biosphere until the potential radiation hazard has been reduced by the natural radioactive decay, and reducing the potential for military use of the weapon-usable plutonium in these materials. The repository is a set of shafts and corridors mined deep into rock in a suitable environment so that water will not bring the radioactivity to the surface. Reactor fuel rods in heavy steel canisters are suitable for such disposal, as is the radioactive material separated by reprocessing—in the form of glass cast inside stainless steel logs weighing as much as a ton or so. The Cheney report looks favorably on reprocessing of spent fuel from U.S. power reactors; a recent study by an expert committee of the National Academy of Sciences finds no merit or necessity in such an approach—certainly not in reprocessing to obtain plutonium to reduce by 20 percent the uranium consumption of the usual U.S. reactor.

Even if the entire world's energy needs were met by nuclear power, the four billion tons of uranium naturally present in the oceans would fuel the reactors for thousands of years. If no new source of energy such as nuclear fusion were perfected in a thousand years, breeder reactors could be used to power the world's energy needs for 500,000 years or more.

Three novel observations in support of nuclear power: First, in terms of global warming, nuclear energy is a far better choice than fossil fuels. For example, burning coal to generate electricity releases carbon dioxide into the atmosphere, which can remain there for more than forty years. This excess carbon dioxide enhances greenhouse warming, each year contributing as much heat to the Earth's surface as was initially generated by the plant! Nuclear power plants, which generate no carbon dioxide, would contribute less to the greenhouse effect over forty years than a coal-burning plant would in a single year.

Second, contrary to popular opinion, the world faces no shortage of its primary nuclear fuel, uranium metal. The world's oceans contain more than four billion tons of uranium, 1,000 times the existing high-grade terrestrial resources. Furthermore, extracting this material would be economically feasible. We encourage the current administration to perform further research into sea water uranium and to determine the exact cost of a large-scale program.

Finally, we recount and support recent developments in commercial reactor technology. Even if, as we argue, existing reactors have reached adequate safety thresholds since the days of Three Mile Island, we encourage aggressive development of and investment in cheaper, safer, and more efficient generations that produce less waste. International commercial cooperation has resulted in novel reactor designs, which offer stronger protection against accidental contamination or meltdown. We encourage the current administration to add federal support to these projects; the safety and environmental benefits of nuclear power will not materialize unless its capital cost is reduced by a factor of two or more. To us, the change in Russian law in 2001 to allow the influx of spent commercial reactor fuel for disposition in Russia is a favorable innovation. But we would prefer to see the client have a choice as to whether the fuel is directly disposed of (without reprocessing) into the mined geologic repository or whether it first has the plutonium removed. The second option is far more costly if new plants must be built, and does not significantly ease the disposal problem—if at all.

Despite our support for nuclear power, we encourage the current administration to accept our analysis that it comes—whether through routine operation or by an unexpected disaster—with a quantifiable cost in human lives. We estimate 35,000 people worldwide will die prematurely from the aftereffects of Chernobyl. The damage, however, is less than that imposed by the coal-fired plants that currently provide half of all U.S. electricity. Coal plants emit sulfur oxide, responsible for much of acid rain and lung disorders; heavy metals such as mercury; and radioactivity present in the three million tons per year consumed annually by each large coal-fired power plant. About 5 percent of this coal ash is used for making concrete for dwellings; on the same basis that leads to the estimate of premature deaths from the one significant civilian nuclear disaster at Chernobyl, 2,000 people die each year from radiation in their homes from the ash contained in concrete.

NATIONAL MISSILE DEFENSE

The most striking characteristics of current threats to U.S. national security are their sheer diversity. For example, surpassing the threat of attack from "rogue" nuclear states, America faces the substantial—if unintentional—risk of bombardment from Russia. Six thousand ex-Soviet nuclear warheads currently remain on high alert, standing by to be launched on minutes' notice. Russian leaders depend on a decaying early warning network for crucial information about an ongoing attack. Malfunctions and gaps in this system, increasingly common byproducts of Russia's shrunken military budgets, increase the possibility of an accidental or unauthorized nuclear launch. Amazingly, these thousands of warheads, poised on a dubious hair trigger and carrying enough combined explosive power to destroy both the United States and Russia, are not seen as a major threat.

In our view, rogue-state ballistic missiles—the centerpiece of current discussions about missile defense—represent a less urgent threat than existing Russian stocks. In early 1998, Garwin served on the nine-member Rumsfeld Commission to Assess the Ballistic Missile Threat to the United States. The final classified report, released also as an unclassified, public Executive Summary,[2] identified three rogue nations that could pose future threats to the United States. In the judgment of the committee, North Korea, Iraq, or Iran might build, within five years, crude intercontinental ballistic missiles (ICBMs) capable of striking U.S. territory. These missiles could carry nuclear or biological warheads, which had been developed independently, purchased, or stolen from another state. The Bush administration and Secretary of Defense Donald Rumsfeld set defense against this form of attack as their first priority—despite the fact that the Rumsfeld report of 1998 specifically identified the short-range ship-launched missile threat against U.S. coastal cities as sooner, more accurate, and easier to mount than would be an ICBM threat. It is the nuclear explosive or the biological warfare agent (say, anthrax), which causes the damage; would that it could be delivered only by ICBM and not by aircraft, or ship, or short-range missiles!

Current military budgets favor the continuation of a Clinton-era program, in which interceptor rockets would be launched from American soil. After satellite detection of a missile launch, rockets would attempt to intercept the incoming warheads midcourse, for example, in the vacuum of space. Under existing plans, the interceptor would detect the warhead thermally and then collide with it in space, destroying the nuclear payload. In our view, such a program is so tenuous that it would require full cooperation from the attacking nation—in the form of shared information about the ICBM's design and possible countermeasures—in order to succeed.

Furthermore, as has been argued for years inside and outside the government, a nation with the resources and technical skill to build ICBMs and nuclear or biological warheads could also devise effective countermeasures. One such possibility would be to divide the payload into dozens or hundreds of smaller anthrax-loaded bomblets, which would fall separately through space to spread their deadly germs over an entire city. Current plans for interception, based on a few-to-one ratio between interceptors and attackers, could not contend with such numerous targets. Another effective countermeasure would enclose the nuclear warhead in a large balloon; the interceptor would collide harmlessly with the balloon, failing to destroy the much smaller warhead inside. Or, the warhead could be placed in a small balloon accompanied by identical empty decoy balloons. The interceptor would have no way of discriminating between harmless empty balloons and the one containing the warhead. Our book discusses these problems in some detail; further information is available in an archive of Garwin's papers, at http://www.fas.org/rlg.

As a more practical alternative to midphase interception, we advocate attacking the missile in its boost phase, when countermeasures are far more difficult. Interceptors could be launched from land or sea, within 1,000 kilometers of the ballistic missile launch site. In the case of North Korea, the United States could use ships located in the Sea of Japan or, operating jointly with Russia, land-based missiles south of Vladivostok. Iraqi ICBMs could be intercepted in boost phase from a single base in Turkey, a NATO member state. Such a defense could not, however, protect against ICBMs from Russia or from China. Both countries are geographically far too large for short-range, land- or sea-based interceptors to reach boost-phase ICBMs.

Tellingly, there is little military enthusiasm for the Bush administration's pet project, defense against the rogue-state ICBM threat. The threat of attack by ballistic missile is modest compared to other means of delivering nuclear or biological weapons. Bloated expenditures on midcourse national missile defense threaten to exclude other programs from defense budgets, reducing the nation's ability to counter more serious threats. The only way to address genuine U.S. security needs is to terminate expensive, ineffective programs like midcourse intercept in favor of a proposal with a reasonable chance of success: boost-phase intercept. A ray of hope: In 2002 the Defense Science Board recommended sea-based boost-phase intercept as a priority.

RUSSIA, CHINA, AND THE ANTIBALLISTIC MISSILE TREATY

Treaties are a tool toward security, not an end in themselves. But the same can be said of contracts in business. The 1972 U.S.–Soviet ABM (Antiballistic Missile) Treaty specifically allows its lapse after six months' warning by one party that the continuation of the treaty would imperil its "supreme national interest." China is not a party to the ABM Treaty, but its actions consequent to the scrapping of the treaty should be considered in deciding on the course of action. The Canadian–U.S. Rush-Bagot Treaty limits naval armaments on the Great Lakes to four warships below one hundred tons. The treaty is still in force but largely ignored as a matter of interpretation. It is clear that for many in the Bush administration, the elimination of the ABM Treaty is a long-sought goal in itself—irrespective of its security impact. We judge that it would be a simple matter to obtain Russian accord on an interpretation or modification of the ABM Treaty (a protocol to the treaty, which is like a codicil to a will), which interprets the ban on deploying a defense of the national territory against strategic ballistic missiles in flight trajectory as referring to strategic ballistic missiles (of a party to the treaty) in flight trajectory. Another option is to add to the list of permitted ABM systems (one for each side) specific agreed systems. As indicated in our book, these would include ABM systems jointly operated by Russia and the United States, and in addition, specific boost-phase intercept systems deployed on land or sea near Korea, Iraq, and perhaps Iran.

As of August 2001, Russia complains that the Bush administration in all its discussions with the Putin government has never indicated specifically what it wants to build. The course of negotiations is never easy to predict, but there is reason to believe that Russia would endorse certain programs such as boost-phase intercept, which could provide effective defense against what is claimed to be the design threat—ICBMs in the hands of one of the three rogue states. Alas, President Bush abandoned the ABM Treaty in December 2001, and it lapsed six months later.

China has been satisfied for years to have some twenty ICBMs, which can reach the continental United States with massive nuclear warheads. These missiles are in silos and could be destroyed by U.S. missiles; hence China has long had development programs to replace or supplement the fixed-site ICBMs with mobile missiles that cannot be destroyed preemptively by a U.S. strike. But there is little urgency visible in this program. Since an avowed goal of many of the supporters of nuclear missile defense (NMD) deployment is protection against the Chinese missile threat, it would be difficult for China to ignore a U.S. NMD system. At times Chinese officials have indicated that if the U.S. NMD had 200 interceptor missiles, China would build 220 mobile ICBMs. In addition, we judge (as did the CIA in a 1999 National Intelligence Estimate) that China would have many countermeasures against an NMD such as the midcourse system now under development.

The terrorist attacks of September 11, 2001, which destroyed the World Trade Center and damaged the Pentagon, killing almost 3,000 people, belie the argument that our foes are too inept to fashion simple countermeasures such as anthrax bomblets or an enclosing balloon. And the much-heard argument, pre-9/11, that we have an annual operating budget of $11 billion to guard our borders against unwanted people or goods (but zero operating homeland missile defense) is shown as sophistry. It is not the *input* of money but the *output* of security that is the measure of value. Analysis should replace debating points; 3,000 dead— exceeding the number who died at Pearl Harbor—is a terrible toll, but only a tiny fraction of the number likely to die if we value ideology above security.

REDUCING THE SECURITY THREAT FROM EXCESS RUSSIAN NUCLEAR MATERIALS

Loose, weapon-usable nuclear materials represent an enormous threat to U.S. and global security. Under existing arms reduction treaties, the United States and Russia will soon have each removed fifty tons of plutonium from disassembled nuclear weapons. In the wrong hands, a tiny fraction of this plutonium could facilitate the rapid construction of illicit new arms. The metal is more portable and more difficult to account for than a discrete warhead. Further, although the warheads themselves are afforded at least a minimal level of protection, the raw plutonium is wrongly perceived to be less attractive to thieves or proliferators. In light of existing flaws in the Russian program for monitoring and guarding nuclear materials, loose plutonium represents an unprecedented international risk.

It takes less than six kilograms of weapons-grade plutonium to make a primitive nuclear explosive. One such device, the bomb dropped on Nagasaki during World War II, killed more than 100,000 people. Weapon designs from the 1950s, more portable and easily deliverable than their wartime predecessors, give greater explosive yields with less plutonium. In a near-worst-case scenario, 50 tons of excess Russian plutonium could yield 10,000 new nuclear weapons. The worst case, paradoxically, may be that one ton would be used to make 200 nuclear bombs.

Excess uranium from disassembled nuclear warheads is easier to use than plutonium for military purposes. Over the next fifteen years, the United States will take delivery of the remainder of 500 tons of highly enriched military uranium from Russia, blended to low enrichment for use in U.S. commercial power reactors. The simplest fission device, like the bomb dropped on Hiroshima in 1945, requires about 60 kilograms of highly enriched uranium; the 500 tons under contract could be converted into 8,000 such weapons. Using more advanced designs, the same amount could yield more than 20,000 explosives.

A joint U.S.–Russian Commission on the Disposition of Excess Weapons Plutonium, on which I served in 1996 and 1997, adopted a dual-track approach to the plutonium problem. One track, known as vitrification, would render excess plutonium more secure by mixing it with highly radioactive nuclear wastes in a steel-encased glass ingot and burying it in the Yucca Mountain repository. In the other track, weapons-grade plutonium would be converted to fuel for U.S. nuclear reactors. Under normal operating conditions, plutonium fuel is mixed with highly radioactive nuclear wastes and could then be buried in the same repository. Although Russia has shown little interest in vitrification, it has plans to use excess plutonium in its own reactors.

Cost, however, is a major factor. Excess weapons-grade uranium can be diluted with natural uranium for use as ordinary reactor fuel, yielding a net commercial profit. However, it costs substantially more to convert warhead-grade plutonium into reactor fuel than to mine, purify, enrich, and prepare uranium from scratch. Therefore, the private-sector corporations involved in nuclear energy would need government subsidies. Having (properly) rejected a $6 billion proposal to convert excess plutonium into reactor fuel, the Bush administration is unwilling to consider the cheaper vitrification approach. At the same time, Russia is determined to keep its weapons-grade plutonium for use in breeder reactors, thereby producing even more excess plutonium.

It lies clearly in U.S. and allied interests to reduce, as quickly as possible, the availability of Russian weapons-grade plutonium. Economically, strategically, and politically, the United States must assume the leadership role in any nonproliferation regime, although its economic allies can be expected to do their share.

The Bush administration is ready to spend more than $100 billion on a ballistic missile shield. In our opinion, existing loose nuclear materials represent a much greater threat to national security than potential rogue missiles. We judge it far more responsible for the Bush administration to review its priorities and support—at a substantially lower price tag—the prophylactic measure of Cooperative Threat Reduction.[3]

NOTES

1. *National Energy Policy: Report of the National Energy Policy Development Group*, May 2001. For sale by the superintendent of documents, U.S. Government Printing Office ISBNO-16-050814-2. It is also available at http://www.whitehouse.gov/energy/National-Energy-Policy.pdf.

2. B. M. Blechman, Lee Butler, R. L. Garwin, W. R. Graham, W. Schneider, L. D. Welch, P. D. Wolfowitz, R. J. Woolsey, and D. H. Rumsfield. *Executive Summary of the Report of the Commission to Assess the Ballistic Missile Threat to the United States*. July 15, 1998, available at http://www.fas.org/rlg.

3. These reflect positions taken in our book, *Megawatts and Megatons: A Turning Point in the Nuclear Age?* (New York: Alfred A. Knopf, 2001). Additional material, such as that provided here, is available at http://www.aaknopf.com/authors/garwin/. In December 2002, the University of Chicago Press published the book as *Megawatts and Megatons: The Future of Nuclear Power and Nuclear Weapons*.

THE DARI

A Unit of Measure Suitable to the Practical Appreciation of the Effect of Low Doses of Ionizing Radiation

Georges Charpak and Richard L. Garwin

A clearer understanding by a wider public of the health effects of radioactive materials arising in the nuclear industry is essential if the public interest is to be served. Even clear and continuous information provided to the public about radiation dose from industry is inadequate to an intuitive and correct understanding of relative risk—in part because radiation exposure is expressed in units that nonspecialists find difficult to comprehend.

We propose the establishment of a unit of irradiation dose to the individual that is equal to that provided to a human being by the naturally occurring radioactivity of human tissue: the *DARI*, from the French for "Dose Annuelle due aux Radiations Internes"—annual dose from internal radioactivity.

To the extent of 90 percent, this radiation is due to potassium 40 (K^{40}), with a half-life of 1.3 billion years, which was present in the cosmic dust from which the Earth was formed about 4.5 billion years ago.

The DARI amounts to less than 10 percent of the natural radiation to which the body is subject, arising from external irradiation from rocks and from cosmic rays. The use of this unit for expressing the individual's radiation dose from an incident or an accident involving radioactive materials would facilitate a proper judgment of its impact, and would avoid unwarranted concerns.

INTRODUCTION

In the physical sciences, an important role is played by units of measure that permit an easy appreciation of the order of magnitude of the items measured. The meter, the kilogram, and the second were thus adopted in engineering. In fields in which the order of magnitude of the object studied is much smaller or much larger, it has been necessary to introduce auxiliary units adapted to daily practice. It is thus in astronomy or in microscopy, where one uses the light-year or the angstrom and the nanometer, respectively.

Over the last few decades there has been a massive increase in use of ionizing radiation. There has been a corresponding evolution of units for the measurement of the intensity of sources emitting radiation and for evaluating their effect on humans.

The extreme sensitivity of instruments to measure radioactivity, detecting even the disintegration of a single atom, has led practitioners, in their particular fields, to deal with figures that have a substantial number of zeros. Thus the becquerel (Bq) is the intensity of a source in which one atom on the average disintegrates each second; it is used for weak sources. The megacurie is often used by an engineer dealing with nuclear wastes; the curie (Ci) being the intensity of a source in which thirty-seven billion atoms disintegrate each second—hence 3.7×10^{10} Bq.

Evaluating the effect of radiation on the human body involves more complex questions. The nature of the ionizing radiation— alpha particles, betas, gammas, heavy ions—makes it necessary to take into account the great variation in the deposit of energy along the trajectory of the particle and on its effectiveness in disturbing the genetic material of living cells. The mechanisms involved in these effects are still a matter of debate, and whether or not a threshold for harm exists gives rise to argument.

CURRENT UNITS OF IRRADIATION

The accepted units (employed by various individuals and groups who must make decisions regarding acceptable doses of radiation), both for the public at large and for workers in the field, are based on the energy deposited by the ionizing radiation. But these units are hardly helpful in providing an intuitive estimate of the nature and hazards of radiation. In this regard, let us just note that the lethal dose of radiation for a human involves the deposition of energy that raises the temperature of the body by a mere thousandth of a degree.

The gray (Gy) corresponds to the deposition of 1 joule per kilogram of living tissue. One takes into account the different sensitivity of the various human organs by weighting the deposit of energy by a coefficient of effectiveness, which leads to the definition of the sievert (Sv). Finally, although it is a matter of reasonable agreement that the lethal dose is four or five Sv for a human being, regulatory authorities interpret the available facts on the induction of lethal cancer by low doses of radiation as giving a probability of cancer linear with the dose, without a threshold of harm for a human being. Specifically, that estimate is 0.04 lethal cancers per Sv of whole body irradiation. Such a coefficient permits the prescription of permissible levels of irradiation to have an acceptable risk for the various populations when weighed against benefits to the economy or to the public health in the use of radiation, for instance, for the diagnosis or treatment of disease.

PROPOSAL FOR A NEW UNIT LINKED TO THE IRRADIATION OF THE HUMAN BODY BY ITS OWN NATURAL RADIOACTIVITY: THE DARI

We suggest an approach that may lead to a more intuitive estimation of the impact of low doses of radiation. The unit that we propose provides a much more immediate idea of the risks run by a given level of irradiation. It is as rigorous as the units used thus far, to which it can be related in a precise fashion. We suggest the DARI. This is close to the irradiation experienced during a single year by an individual, due to the radiation emitted by the radioactive materials present in the human body that have nothing to do with any line of work. The DARI is to be defined as 0.2 millisieverts, precisely, although the annual dose itself is about 10 percent less.

The two principal radioactive substances that contribute to this internal irradiation are K^{40} (a natural isotope that is a permanent component of living tissue) and C^{14}, produced in the air by cosmic rays and present in all living organisms.

We prefer this standard of internal radiation to one representing the average human irradiation from all natural sources—at a level about ten times larger. This total varies too much with geography and altitude to serve as a reasonable standard.

Our interest in a standard such as the DARI arises from the fact that even much weaker irradiations than this internal exposure give rise to futile controversy and can distort crucial choices such as those that concern energy supply for centuries to come.

First we discuss more extensively the origin of this internal irradiation, which will permit us to estimate readily the scale of certain incidents or accidents related to nuclear power.

NATURAL SOURCES OF INTERNAL AND EXTERNAL IRRADIATION

Our planet was formed by the aggregation of the dust from dead stars. All of the chemical elements that compose Earth were synthesized through the nuclear reactions that take place throughout the life of such stars. Certain elements are radioactive with a mean life in hundreds of millions or billions of years and are always present.

Uranium, thorium, and potassium play an important role in creating the molten core of Earth. The energy of their radiations has melted and maintains the molten sphere of iron-nickel, some 3,500 km in radius, which constitutes the core of our planet.

These primordial radioactive wastes are present everywhere

among us. K^{40}, which is a natural isotope of potassium, has a half-life of 1.3 billion years. It pervades living organisms, and the human body of 70 kg mass is host to about 6,000 Bq—that is, 6,000 disintegrations per second.

Uranium is widely distributed on Earth and constitutes about three millionths of Earth's crust. Its presence in rocks contributes a significant portion of the natural irradiation to human beings. It contributes also by the radioactivity of one element in its decay chain—radon 222 (of half-life 3.8 days), which is a noble gas emitting alpha particles. Radon and its radioactive daughters are responsible for more than half of the natural irradiation of humans; the U.S. government recommends effective ventilation of homes in order to reduce the level of radon from building materials or from seepage from the underlying rock.

NUCLEAR ENERGY AND PUBLIC HEALTH

Uranium has a special importance since the discovery of nuclear energy. By use of uranium fission in a nuclear reactor, as much energy could be extracted from the uranium of any given portion of Earth's crust as if it were pure coal.

Of course, because of the cost of extraction of uranium from such lean deposits, it is currently obtained from the much smaller deposits of higher grade ore—ranging from 0.1 percent to 14 percent uranium by weight. But studies show that it is possible to obtain uranium from seawater (where it constitutes only about three billionths by weight) with a cost that is a mere fifteen to fifty times larger at the moment than that required to obtain uranium from the deposits that now serve to feed nuclear power plants. Uranium from seawater thus gives an affordable and practically unlimited supply of fuel for fission power.

In the future, which in the more or less long term is threatened by the exhaustion of the energy resources based on fossil fuel, it is thus legitimate to consider the major role that can be played by nuclear energy. This is particularly true considering the enhanced greenhouse effect from the carbon liberated to the atmosphere by the normal combustion of coal, oil, or even natural gas. However, the enormous amount of radioactivity produced in the course of releasing energy from fission raises inevitable questions and anxieties about the danger for our generation and those to come of the massive deployment of nuclear power.

The degree to which nuclear power is accepted depends upon the resources of the individual countries, but also upon a realistic evaluation of the dangers and problems the various alternative energy sources present to the human species. Decisions are rendered more difficult by the sometimes irrational character of the debates concerning the effects on humans of ionizing radiation of various origins, to which we are exposed, voluntarily or not.

A major problem right now for the nuclear industry is to show that it is capable of caring for radioactive waste from nuclear power plants in a satisfactory fashion for generations to come, and is able to practically eliminate the possibility of a major catastrophe like that at Chernobyl in 1986.

To make a proper judgment of various proposed courses of action, it is helpful to take into account the irradiation to which humans are exposed independently of nuclear energy.

RELATIVE IMPORTANCE OF NATURAL SOURCES OF IRRADIATION, INTERNAL OR EXTERNAL

Radiation from internal sources is an absolute floor below which it is impossible to sink. Natural radiation overall, about ten times larger, represents a level to which we should refer to evaluate the limits imposed on exposure from the nuclear industry, and in order to evaluate the seriousness of incidents or accidents. We will now examine several sources contributing to natural radiation, using as the unit of irradiation the sievert or the thousandth of a sievert—millisievert (mSv). Cosmic rays shower Earth and provoke nuclear reactions in the upper atmosphere caused by the energetic protons of the cosmic rays. At sea level the cosmic rays contribute an annual irradiation about 0.5 mSv. The intensity increases with altitude, which contributes more irradiation in a year to airline crews than is legally permitted for workers in the nuclear industry—50 mSv per year.

In the atmosphere, cosmic rays transmute nitrogen into radioactive carbon—C^{14} with a mean life of 5,000 years. C^{14} is present in the form of carbon dioxide within the atmosphere. Incorporated by plants into their tissues, C^{14} pervades all living things and, along with K^{40}, present since the beginning of Earth, contributes the most important and least avoidable part of internal radiation. For a person of 70 kg weight, C^{14} contributes about 4,000 becquerel; together with the 6,000 Bq from potassium 40, one has thus an activity of 10,000 Bq in the average human being. Taking into account the relative biological effectiveness of this radiation on different organs, this source of 10,000 Bq contributes about 0.17 mSv per year. Incidentally, it is the same 0.17 mSv per year for a child and for an adult of whatever weight, since it is the energy deposited per kg that is measured by the Sv—1 joule per kg.

Now we compare this level with the total irradiation experienced by humans from other natural sources, totaling, as indicated, about 2 mSv per year. One needs to take into account the radioactivity of the soil containing more or less radioactive material—notably potassium and uranium. In France, there is a variability of a factor 3 from 1 mSv per year in the environs of Paris to 3 mSv in Brittany. On our planet, there are vast populated regions where the natural radioactivity is much greater. Added to the exposure from rocks, there is the radioactivity due to radon, and the cosmic ray intensity that varies with altitude.

Natural irradiation of the average U.S. citizen, like that of the French, amounts to about 2.5 mSv per year, to which must be added the irradiation due to medical diagnostics—on the order of 1 mSv per year. The variation in natural radioactivity within France, about 1 mSv per year, exceeds the limit imposed by the law on the exposure of the civil population by the nuclear industry. At this level, it has not been possible thus far to demonstrate any impact on public health; there is no direct evidence of harm caused by irradiation of this magnitude.

It seems to us instructive to choose as a practical unit of irradiation close to that due to the unavoidable K^{40} and C^{14} in the body, amounting to 0.17 mSv per year, which is quite uniform among the world's population. We have rounded this to 0.20 mSv per year and call it the DARI. According to the International Commission on Radiation Protection (ICRP), exposure to a DARI conveys a probability of incurring lethal cancer of ten parts in one million. If a lethal cancer corresponds to twenty years of life shortening, each DARI then costs the individual one hour of life expectancy. This calculation is rejected by some who believe in the existence of a threshold, below which there is no harm from radiation. In either case, one hour per year is little enough to pay for the gift of a body inherited from the stars and the cosmic rays.

The natural variability from place to place amounts to five DARI or more. Nevertheless, we should not countenance the addition of even one DARI to the individual's radiation burden without considering the benefits to that individual and to society. Table 1 demonstrates the relative importance of various widespread sources of irradiation.

The exposure of 250 DARI imposed on a worker in the nuclear industry corresponds to a reduction in life expectancy (250 hours) equal to that produced by the smoking of five cigarettes per day during that same year.

This risk of cancer to a worker receiving the maximum permitted dose in the nuclear industry should be compared with the occupational risks associated with other industrial or commercial activities. For example, driving an automobile in traffic exposes the driver to carcinogenic exhaust fumes (especially particulate matter) with a greater risk of cancer.

Recently one has seen in France polemics over accidental releases of radiation whose impact is less than one-hundredth of a DARI, exposing only a local population to this tiny augmentation in irradiation.

The DARI puts into perspective these polemics, of which the impact on the political scene and in the media is disproportionately large compared with the substance. The debate should center on the following problems as regards future energy:

Table 1. The Relative Importance of
Various Widespread Sources of Irradiation

0.1 DARI	Dose received in France from the nuclear power industry
5 DARI	The soil in the environs of Paris
5 DARI	Cosmic rays at sea level (increase of 1 DARI per 50 m at altitude)
5 DARI	Average diagnostic radiography*
5 DARI	Limit established for the irradiation of the public by the nuclear industry
10 DARI	The soil in Brittany
40 DARI	Single CAT scan
250 DARI	Annual maximum dose for a worker in the nuclear industry
25,000 DARI	Lethal dose for the average human
300,000 to 500,000 DARI	Dose delivered as local irradiation to treat a cancer

*The DARI is intended to represent an effective dose. If 2 milligray of gamma radiation is delivered to the left side of the body, the equivalent (and effective) whole-body dose is one millisievert or 5 DARI. According to the linear hypothesis for the effects of low doses of radiation, the same probability of cancer will result as if one millisievert were delivered to the whole body—although any tumor will appear only on the left side of the body. A similar approach applies to the effect of a small diagnostic dose of radioactive iodine—effects of which would be limited to the thyroid but that could be expressed in an equivalent whole-body dose in microsieverts or DARI.

- What are the real and comparative hazards of various sources of energy now available to humanity?
- After the exhaustion of the fossil energy supplies, what are the options?

We are expecting nine billion humans to inhabit the Earth in the middle of this century, compared with six billion at this moment. In the industrialized countries, about 20 percent of the population dies from cancer. Among the sources of cancer, about half have been identified as due to lifestyle—tobacco, alcohol, obesity, diet—and seem avoidable. About 2 percent of cancers are thought to arise from carcinogenic materials used in industry or from automobile exhaust fumes.

One must be alert to reduce to a minimum all of these hazards, especially the more important ones. Those that are due to radioactivity are among the easiest to measure, and they must be maintained at an acceptable level. A unit of measure that takes into account the unavoidable natural self-irradiation of a human being seems appropriate, bearing in mind that it has not been directly demonstrated to have an effect on health.

According to the severe criteria used by regulatory authorities to evaluate the impact of radiation on public health, the DARI shortens life by about one hour per year. And the French accordingly lose six minutes per year because of their dependence on nuclear power—assuredly less than the hazard of the coal alternative. The table shows, however, that there is substantial merit in reducing the exposure to the average individual from diagnostic X-rays, which by the same calculation shorten life by an average of five hours per year.

BUILDING PUBLIC UNDERSTANDING OF SCIENCE

A Matter of Trust

Judith A. Ramaley

In a recent text, *Science in Public*, Jane Gregory and Steve Miller (1998) discuss the fact that scientists and policymakers now insist that "the public must understand science if they are to be useful citizens, capable of functioning correctly as workers, voters, and voters in a technological age." What does public understanding actually mean? In the Science and Engineering Indicators (SEI) published by the National Science Foundation every two years (National Science Board 2000), the implicit answer, based on the data provided, is that public understanding encompasses (1) interest in science and technology and attentiveness to these issues and the recognition that science and technology have both strengths and limitations as human enterprises; (2) understanding of scientific and technical concepts and vocabulary; (3) attitudes toward science and technology policy issues; and (4) use of various sources of scientific and technical information ranging from print media to the Web (National Science Board 2000).

How much do our citizens actually *know* about science? According to the 2000 SEI, about 20 percent of U.S. adults think they are very well informed about new scientific discoveries and about the use of new inventions and technologies, and most adults express an interest in these questions. Only about 25 percent of Americans understand enough about the nature of scientific inquiry to be able to make informed judgments about the scientific basis of results reported in the media. Only about 14 percent of Americans are attentive to science and technology policy issues except when a crisis of some kind draws their concerned attention. Since Americans receive most of their information about science and public policy issues from television news programs and newspapers (National Science Board 2000), it is important to think about both how scientific issues are presented through these media and how well-prepared citizens are to judge critically what they see and hear.

It is also important to think about a relatively new topic, *Web literacy*. Increasingly, as computer technology and Internet access pervade the workplace, public spaces such as libraries and museums, and home environments, more people are trying to find specific information on the Web. In fact, the Web is becoming a kind of public library. The 2000 SEI estimate that approximately 8.8 million people have tried to find some scientific information on the Web, including information on the issues that remain most controversial in this country—nuclear power, genetic engineering, and space exploration. How prepared are our citizens to distinguish valid sources of information from useless or even dangerous misrepresentations of what is known and what is not?

PUBLIC DISTRUST OF SCIENCE

Why do so many people distrust science? A new element that has been added to our concept of scientific literacy is the importance of confidence and trust in science and in scientists and policy-makers who are making decisions about issues that have a scientific or technological basis. Although in calm times 85 percent of Americans think that the world is better off due to science (National Science Board 2000), a real test of confidence comes when people are scared. To understand the dynamics of public response during a scare, we can turn to the United Kingdom, where public understanding of science has been tested by such serious concerns as mad cow disease and genetically engineered foods. In the United Kingdom,

> Science often meets the public in times of crisis. Their relation-
> ship is conducted fleetingly and acutely through mass media,
> which emphasize emotion in place of what are often rather
> scarce *facts*. And when scientists cannot agree on a solution to
> a scientific problem, it is not surprising that the public turns to
> moral or emotional solutions in order to get on with their lives.
> The highly charged environment pushes everyone involved to
> extreme practical measures and to polarized points of view, and
> often results in a breakdown of both trust and communication
> between political and scientific authorities and the publics they
> purport to serve. (Gregory and Miller 2002)

In its report *Science and Society* in 2000, the House of Lords Select Committee on Science and Technology (House of Lords 2000) documents a clear change in the approach of British scientists and government officials toward engagement with the public on issues of science. According to Alan Irwin (2001), the shift in strategy from *informing* the public to *talking* with the public began with the publication of a report from the Royal Commission on Environmental Pollution (RCEP) in 1998, which advocated much greater "transparency and openness within decision-making" and stressed the importance of engaging the public in a meaningful way in exploring policy issues and identifying the values and concerns that should be addressed by policymakers.

In the United Kingdom, the conversation about public understanding of science now focuses on the need for the public to understand the fundamental nature of scientific uncertainty and the ways in which scientific inquiry proceeds. The British have recognized the absolute necessity of attention to public trust and confidence. They have learned that trust cannot be gained simply through providing information about science but by direct dialogue and discussion about the issues. This task is referred to as "replacing the deficit model" (namely the idea that it is necessary only to provide the public with facts and they will understand and accept decisions based on scientific input) with a model of honesty and respect for the interests of citizens and their need for a clear moral and value structure underlying the facts.

As the House of Lords report points out (House of Lords 2000), "knowledge obtained through scientific investigation does not itself have a moral dimension; but the ways in which it is pursued, and the applications to which it may be put, inevitably engage with morality." One could also argue that these moral dimensions have a compelling emotional foundation that must be addressed as well. Many people faced with a crisis ask for comfort and for the relief of their fears, not for the facts. The Select Committee suggests "by declaring the values which underpin their work, and by engaging with the values and attitudes of the public, scientists are far more likely to command public support and engender trust. These attitudes and values must be weighed along with scientific findings by policymakers."

Direct engagement with citizens was incorporated into the Citizen Foresight project launched by the London Centre for Governance, Innovation, and Science to address the future of agriculture and the food system (Irwin 2001). Twelve British citizens selected at random came together for ten weekly meetings to listen to evidence, ask questions, and draw their own conclusions. The lay panel members then chose the particular topics they wished to explore further, invited expert witnesses to address these questions, and then drew up their own conclusions. The panel explored not only the facts but also the deeper ethical and emotional issues associated with questions about food supplies and agricultural technologies. The Citizen Foresight project demonstrates that scientific knowledge must be grounded in a moral and ethical foundation that is seen as legitimate by the public and accepted as responsive to their needs and interests.

EDUCATIONAL STRATEGIES TO BUILD UNDERSTANDING IN SCIENCE

What educational strategies might we employ to build an understanding of science that is emotionally valid and trusted as well as useful? In recent years, we have grown accustomed to talking about the importance of promoting a public understanding of science through how science is taught in the schools. In a recent report in *Physics Today,* Lopez and Schultz (2001) argued that two revolutions have changed the approach to science education in grades K–8. One revolution is in the goals of science education (the idea that science is for all Americans), the other is a revolution in how science is taught (science should be something that students do, not something done to them). As they put it, "the idea that science education should be for all children, not just the best and brightest, reflects a recent, fundamental change in the relationship between science and society" (Lopez and Schultz 2001).

Project 2061 of the American Association for the Advancement of Science is often cited as the milestone for this new thinking about the role of science and society. *Science for All Americans* declared that it is possible to construct "a common core of learning in science, mathematics and technology for all young people, regardless of their social circumstances and career aspirations" (American Association for the Advancement of Science 1990).

Project 2061 levels some serious criticism against the way science is taught in most schools. Although written in 1990, this critique is still frequently true today.

> The present science textbooks and methods of instruction, far from helping, often actually impede progress toward science literacy. They emphasize the learning of answers more than the exploration of questions, memory at the expense of critical thought, bits and pieces of information instead of understandings in context, recitation over argument, reading in lieu of doing. They fail to encourage students to work together, to share ideas and information freely with each other, or to use modern instruments to extend their intellectual capabilities. (American Association for the Advancement of Science 1990)

No wonder so many adults remember science as something to be avoided at all costs and often fail to see that science is connected in any meaningful way to their own lives and interests. It is also no surprise that in times of crisis and fear, adults who learned their science this way would turn to spiritual advisors and rely on rumors, no matter how unfounded, rather than seek out scientific information or turn to scientists for guidance and advice.

The idea that science is for everybody is a return to a philosophy that infused public education in the late nineteenth century. According to Diane Ravitch (2000), in that era, the schools were being called upon to educate all children to high standards. Educators of that era believed that the best way to improve society was to offer a sound education to as many people as possible. During a prolonged period of progressive education during the twentieth century, reformers worked to make the schools more practical, egalitarian, and utilitarian. Throughout much of the first half of the twentieth century, young people entered the workforce immediately after eighth grade or after high school. Reformers argued that they did not need to discipline and train their minds with Latin or science or history. They needed to learn practical skills. Ravitch argues that the net result of this movement was the development of a strong anti-intellectualism in this country and a distrust of experts of all kinds. The nation moved away from a strong academic curriculum, except for those few young men— and even fewer young women—who were preparing to go on to college. According to Ravitch (2000), "whenever the academic curriculum was diluted or minimized, large numbers of children were pushed through the school system without benefit of a real education. As the academic curriculum lost its importance as the central focus of the public school system, the schools lost their anchor, their sense of mission, their intense moral commitment to the intellectual development of each child."

Now we have come around full circle to the philosophy of educators from a century ago. As Project 2061 declares, "America's future—its ability to create a truly just society, to sustain its economic vitality, and to remain secure in a world torn by hostilities—depends more than ever on the character and quality of the education that the nation provides for its children" (American Association for the Advancement of Science 1990). What kind of curriculum can promote scientific literacy? Many would argue that the appropriate curriculum includes "the systematic study of language and literature, science and mathematics, history, the arts and foreign languages . . . in order to convey important knowledge and skills, cultivate aesthetic imagination, and teach students to think critically and reflectively about the world in which they live" (Ravitch 2000) and to help students learn to act in an ethical and principled fashion, mindful of their responsibilities to themselves and to others.

What will it take to prepare all young people for lives of citizenship and social responsibility as well as success in a workplace increasingly shaped by science and technology? It has been argued that scientific thinking is a rare and often counterintuitive experience. Robert McCauley (n.d.) suggests that "science requires forms of thought that human beings find extremely hard to master." For McCauley, science is a product, not an experience, the collective work of a community of scholars "for whom prestige, fame and wealth turn, in no small part, on their seizing opportunities to correct one another's theories and observations." He makes the case that "such communal features of the scientific enterprise establish and sustain norms that govern scientific practice and ensure in the long run that the collective outcome of the efforts of mistake-prone individuals is more reliable than any of their individual efforts in the short run."

In his book *Historical Thinking,* Sam Wineburg (2001) offers a different interpretation of the problem of public understanding. He makes the case that "historical thinking, in its deepest forms, is neither a natural process nor something that springs automatically from psychological development." He argues that it is much easier to memorize facts, dates, and names of historical figures than it is "to change the basic mental structures we use to grasp the meaning of the past" (Wineburg 2001). In his hands, history becomes an example of the challenge of any form of disciplined thinking, any process of seeking to get beyond the surface of a subject to its underlying warrants for truth. The similarity between Wineburg's arguments and McCauley's leads to the suggestion that the public is unlikely to understand any complex thinking unless during formal education they acquire a deeper understanding of the ways of knowing that field. As Wineburg puts it, "new information is often no match for deeply held beliefs" (Wineburg 2001). If during our education, we are never required to examine those deeper assumptions, acquired early and applied without thought to the challenges of daily life, we will not be responsive to the insights and knowledge generated by any discipline, including the sciences and mathematics.

Wineburg quotes Carl Becker's essay "Everyman His Own Historian" at length. The basic message is that we are all called upon to engage in historical thinking every day—to see the underlying motives of the authors of the texts we read, to mine truth from the quicksand of innuendo—half truth and falsehood that threatens to engulf us—to live with the knowledge that certainty must always elude us (Wineburg 2001). By analogy, we are also all our own scientist. The study of history or the study of science can teach us to think and reason in sophisticated ways, but achieving this aim is not easy.

Students who never *do* any science but simply *read* about it or *listen* to lectures are likely to acquire a sense of certainty about what is known and a false impression about what science is and how scientific knowledge is attained. People who think that science is a product rather than a messy process of inquiry can become profoundly uncomfortable when they are brought face-to-face with the uncertainties and arguments at the frontiers of science. This often happens when they most want to have clear simple answers to emotionally laden questions. At such times, people may prefer the opinions of their friends or trusted advisors over the information provided by scientists, especially when scientists are deeply divided over an issue, as they often are.

Students can best learn how science is done by doing it. It is important to (1) provide genuine experiences of doing science throughout the school years from preschool through undergraduate education; (2) link the questions addressed by scientific means to issues that students care about; and (3) integrate scientific exploration with other disciplines so that all students can see how science contributes to understanding in any field and, of equal importance, how other fields contribute to science. When science is meaningfully connected to things that young people care about, it becomes an experience rather than a product to be memorized.

INFORMAL EDUCATION FOR ADULTS

According to Rachel Young (2000), "the crux of the science literacy problem is that, without the tools to assess the merits of various claims of scientific truth, the public may be unable to distinguish revolutionary science from sheer quackery." In the most recent SEI, 36 percent of the respondents thought that astrology was either *very* or *sort of* scientific. One solution often proposed for dealing with this shocking lack of understanding about what science is and what it isn't is to increase both the quality of science reporting and the number of stories appearing in the press. According to a recent report from the Pew Research Center for the People and the Press cited by Rachel Young, only 39 of the 689 most closely followed news stories of the past fifteen years had anything to do with science or medicine (Young 2000). Most of the stories that did have scientific content were about disasters, natural or manmade. In fact, of some 1,700 daily newspapers in this country, only 30 at most cover science routinely, and the volume of that coverage is decreasing (National Science Board 2000). However, there is very little evidence that exposure to information per se can lead to either deeper understanding or an ability to incorporate scientific knowledge into better decision making.

As knowledge gained from educational experiences fades, there must be continuing ways for adults to learn about progress in science and to experience the realities of scientific inquiry. One way is through informal science education. Joel Bloom (1992) describes the concept of informal education as the various ways in which people learn outside the classroom—reading books and magazines, watching television or movies, observing the natural world, visiting a museum. Very little is known about what people learn this way and how well they retain and use what they have learned. However, it is clear that these experiences are direct, personal, and very real. In all of these experiences, people can follow their own interests, browse until they find something really interesting to them, and spend as much time as they like. There is some evidence that learning that has rich emotional content and personal meaning may last longer and be more useful. Given the attention to values in the United Kingdom (House of Lords 2000) and recent discussions about emotional intelligence in the United States (Bloom 1992), this question should be more thoroughly explored.

Many scientists are embracing the importance of outreach to the public in order "to help frame the questions to be posed, provide assessments about current conditions, evaluate the likely consequences of different policy or management options, provide knowledge about the world, and develop new technologies" (Lubchenko 1998), while at the same time communicating the uncertainties of these situations and educating citizens about the issues. While this is certainly to be welcomed over the previous tacit assumption that to talk about one's work was to commit the sin of self-aggrandizement, this approach begs the question of how to ensure that the scientist is talking about something that citizens are actually interested in and reaching the underlying emotional and personal questions that we all have about what these things mean for us and for our families and communities. James Weber (2001) has examined some of the actual things that happen when people talk about difficult subjects that have scientific or technical content. Weber offers some interesting thoughts about how to approach scientific communication. First, it is important to think about communication as a process of mutual interaction and a seeking of understanding, rather than simply as a means to transmit knowledge accurately to the public. Inevitably, in this process information ceases to be value-neutral. It acquires positive or negative connotations depending upon how the participants interpret the information. Second, it is important to put information into a local context. Sweeping generalizations are beyond the grasp of most people who do not think abstractly. The scientists who are communicating cannot easily anticipate how their information will be received. Personal experience often overrides more generalized information, especially when we are dealing in probabilities. Third, once a discussion is opened up, the information will be filtered through the personal experience and emotional needs of the listener. Only real conversation can disclose this process of meaning-making and address the misunderstandings that may result. Furthermore, recent research on learning has shown quite clearly that at any age, new knowledge can only be absorbed and put in context if the participant can uncover older *untrue* knowledge and discard it (Bransford 1998). This also requires careful listening and a thorough exploration of ideas.

SUMMARY

It is now very clear that public understanding of science requires more than simple faithful transmission of knowledge from scientists to the rest of the public. A shared and mutually beneficial interaction or engagement is necessary, both during formal K–12 and undergraduate education and in adulthood in order to generate a sense of trust and to allow for an exploration of deeper beliefs that interfere with the absorption of new information generated by scientists. Most people learn better when they can directly experience the process by which knowledge is generated and can learn for themselves what scientific inquiry is really all about. When science is taught as certain answers to other people's questions, the stage is set for citizens to have difficulty embracing the value of science in their lives. When science is not emotionally satisfying, it will fail to address deeper questions of identity and personal experience and will be rejected in favor of less reliable sources of information and advice.

REFERENCES

American Association for the Advancement of Science. Project 2061. *Science for All Americans*. New York: Oxford University Press, 1990.
Bloom, Joel. "Science and Technology Museums Face the Future." *Museums and the Public Understanding of Science*, ed. John Durant. Science Museum in association with the Committee on the Public Understanding of Science, 1992.

Bransford, John D., Ann L. Brown, and Rodney R. Cocking. *How People Learn: Brain, Mind, Experience and School.* Washington, D.C.: National Academy of Sciences, 1998.

Goleman, Daniel. *Working with Emotional Intelligence.* New York: Bantam Books, 1998.

Gregory, Jane, and Steve Miller. *Science in Public. Communication, Culture and Credibility.* Cambridge, Mass.: Perseus Publishing, 1998.

Gregory, Jane, and Steve Miller. *A Protocol for Science Communication.* Posted at www.ucl.ac.uk/sts/sm/sciencec.htm, 2002.

House of Lords Select Committee on Science and Technology. *Science and Society.* London: The Stationery Office, 2000.

Irwin, Alan. "Constructing the Scientific Citizen: Science and Democracy in the Biosciences." *Public Understanding of Science* 10 (2001): 1–18.

Lopez, Ramon E., and Ted Schultz. "Two Revolutions in K–8 Science Education." *Physics Today* 54 (2001): 44–49.

Lubchenko, Jane. "Entering the Century of the Environment. A New Social Contract for Science." *Science* 279 (1998): 491–97.

McCauley, Robert. *Comparing the Cognitive Foundations of Religion and Science.* Atlanta, Ga.: Emory University Cognition Project, Report #37, n.d.

National Science Board. *Science and Engineering Indicators.* Washington, D.C.: U.S. Government Printing Office, 2000.

Ravitch, Diane. *Left Back. A Century of Battles over School Reform.* New York: Simon and Schuster, 2000.

Weber, James R. "The Communication Process as Evaluative Context: What Do Nonscientists Hear When Scientists Speak?" *Bioscience* 51 (2001): 487–95.

Wineburg, Sam. *Historical Thinking and Other Unnatural Acts.* Philadelphia: Temple University Press, 2001.

Young, Rachel. "Whadda You Know?" In *The Sciences.* New York: New York Academy of Sciences, 2000.

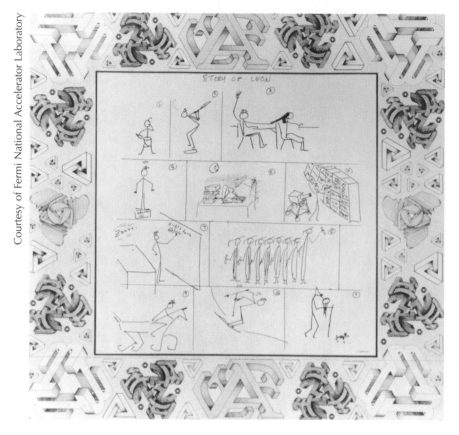

Story of Leon: I distinctly recall my awesome skill at learning the physics that children, age zero to six years, need to negotiate in the complex world, for example, to walk upright and get milk out of a bottle. Fortunately, the necessary chemistry and biology came naturally. Applying the skill later to Little League and other related institutions designed to torture preteenagers was much less successful. It was more fun to dip Alice's blonde hair in the inkwell provided on all our school desks. How else to show undying love? My schoolwork was exemplary but required long hours of study until very late at night. This led to laboratory research. The oscilloscope was my essential companion as we designed electronics for our experiment. Then there was teaching and the pride of a growing number of graduating students. Finally the time came to relax, ride horses, ski, and grow old gracefully.

PART 6
THE LEDERMAN LEGACY FOR EDUCATION

LEON M. LEDERMAN

A Brief Biography

Stephanie Pace Marshall, Judith A. Scheppler, and Michael J. Palmisano

Leon Max Lederman is a person who embodies the joy of living, learning, and giving. He was born on July 15, 1922, in New York City to immigrant parents from the former Soviet Union. While not college-educated themselves, they strongly supported education for their children. Lederman credits his older brother, Paul, as one of the influences on his interest in science. Paul was a "tinkerer of unusual skill" who allowed young Leon to observe him in his basement laboratory in exchange for completing Paul's household chores.

Reading was another force that drew the young Lederman to science. He read such books as Albert Einstein's *The Meaning of Relativity*, *Hunger Fighters* and *Microbe Hunters* by Paul de Kruif, and *Crucibles* by Bernard Jaffe. Lederman found a newspaper article recounting Carl Anderson's Nobel Prize–winning research to be romantic, exciting, and mysterious. These portrayals of science as a romantic detective story caused Lederman to fall in love with science.

Lederman received a B.S. in chemistry from the City College of New York in 1943. Immediately following graduation, he spent three years in the United States Army, eventually achieving the rank of second lieutenant in the Signal Corps. After his discharge, he entered Columbia Uni-

Courtesy of Illinois Mathematics and Science Academy, Brian Quinby

Lederman engaged in conversation with IMSA students in the Grainger Center for Imagination and Inquiry.

versity, not far from where he grew up, to pursue graduate work in physics.

Spending three years away from formal education affected Lederman's study habits. Professors and students alike were all struggling to adjust to academics and life in general after World War II, and Lederman experienced difficulties. Friends at the Massachusetts Institute of Technology (MIT) encouraged him to persist. He considered transferring to MIT, but in the end remained at Columbia. When Columbia built what was at that time the world's largest atom smasher (cyclotron), he was intrigued enough to pursue work in the emerging field of particle physics. He received his M.A. in 1948 and his doctorate in 1951, both from Columbia.

As a graduate student, Lederman exhibited the wit and humor that still characterizes his many public speeches. Columbia University professors thought well of Lederman and his sense of humor and so among his many job offers was one to stay at Columbia. He joined the faculty and learned much from his mentor I. I. Rabi. As a very popular teacher of both science and nonscience majors, Lederman quickly rose to the rank of full professor in 1958. In recognition of his research and administration skills, he became the director of Nevis Laboratories in 1961.

Courtesy of Illinois Mathematics and Science Academy, Brian Quinby

Lederman, as inaugural Resident Scholar in the Great Minds Program, speaks to students about becoming and being a scientist.

In his early years at Columbia, Lederman and his colleagues discovered the long-lived neutral K-meson and confirmed the nonconservation of parity, demonstrating the handedness of the universe. Research in 1962 with Columbia colleagues Jack Steinberger and Melvin Schwartz led to the discovery that distinguished two different types of neutrinos (and won the 1988 Nobel Prize in physics). Attracted by machines of the highest energy, Lederman tirelessly worked at the frontier laboratories of the world where he probed, measured, and observed basic constituents of matter to learn more about the universe. He simultaneously made an impact on national and international science policy through his service and leadership, testifying before Congress and advising funding agencies, universities, and individuals.

Lederman's pursuit of forefront physics led to his involvement with the creation in 1967 of a "truly national laboratory" in Illinois. Guided by his beliefs in providing the best facilities for fundamental discoveries, he advised Robert R. Wilson, the first director of Fermi National Accelerator Laboratory (Fermilab). As head of the Users Group in 1974, Lederman delivered the laboratory's dedication

Fermilab Visual Media Services

Lederman in his office at Fermilab, September 1996.

address. Lederman led an experiment at Fermilab, beginning in 1970, that discovered the bottom quark in 1977.

In 1979 Lederman succeeded Wilson as director of Fermilab. Under Lederman, Fermilab regained its forefront status as the highest energy accelerator in the world, the Tevatron. Hoping to stimulate the breadth of particle physics done at Fermilab, he introduced the idea of the inner space/outer space connection and instituted the first theoretical astrophysics group at a particle physics laboratory.

Lederman receives the 1988 Nobel Prize in physics at the hand of King Gustav of Sweden, December 10, 1988.

Lederman has never been afraid to blaze new directions, take chances, or work with new partners, even astronomers. In 1983, while hiking in the Dolomites with David Schramm, the two conceived the idea of a NASA-funded astrophysics group at Fermilab to exploit the deep connections between elementary particle physics and cosmology. Leon challenged (that's the polite word for it) NASA to be bold enough to fund such an innovative program. NASA responded positively, and Leon recruited Michael Turner and Rocky Kolb to start the group. The interdisciplinary field of particle physics and cosmology has blossomed, advancing our understanding of how the universe began as well as the fundamental laws that govern it. The Fermilab group was so successful that other Department of Energy labs have followed Leon's lead, initiating their own astrophysics groups.

Thinking globally and acting locally, Lederman forged Latin-American collaboration in physics and science and launched exchange programs for countries in need of improved educational systems and up-to-date research facilities. Twenty-eight years of teaching at Columbia instilled in Lederman a deep commitment to science education. He infused this passion for education throughout Fermilab. His scientific and administrative credentials enabled him to find solutions for math and science literacy problems. Dozens of programs developed by Lederman and Fermilab's education arm, the Friends of Fermilab, now reach students and teachers from kindergarten through high school, worldwide.

In his career as a research scientist and science educator, Lederman has successfully conducted forefront research while communicating its value to the public. He has published more than two hundred professional articles, more than one hundred articles for the general public, and three books. He has mentored fifty-one graduate students who have completed their doctorates under his guidance. He is recognized with honorary degrees from more than thirty universities around the world, and his awards include election to the National Academy of Sciences (1965), receipt of the National Medal of Science (1965), the Franklin Institute's Elliott Cresson Medal (1976), the Wolf Prize in Physics (1982), the Nobel Prize in 1988, and the U. S. Department of Energy's Enrico Fermi Prize in 1993.

Lederman's commitment to public service includes serving as science advisor to the governor of Illinois (1989–1993), president of the American Association for the Advancement of Science (1991), NSRC advisory board member (1996 to present), and commissioner of the White House Fellows (1997–2001).

Lederman has long recognized the importance of science education. He was one of the initiators and a chief advocate for the formation of the Illinois Mathematics and Science Academy (IMSA). IMSA was established by the Illinois General Assembly and opened its doors in 1986. Lederman served as a founding trustee until 1998 when he became the inaugural resident scholar of IMSA's Great Minds Program. He is also one of the founders and chairman of the board of the Teachers Academy for Mathematics and Science (TAMS), a university-based, not-for-profit intervention in Chicago and in public schools throughout Illinois. TAMS is devoted to the professional development of primary school teachers in the teaching of science and math.

On the national and international front, Lederman has helped to organize an informal coalition of scientists and educators to reexamine the science curriculum of U.S. high schools in the light of the science standards designed by the National Academy of Sciences and the American Association for the Advancement of Science (AAAS). The project, American Renaissance in Science Education (ARISE), has organized two workshops in 1995 and 1998 addressing issues of science curricula. ARISE is an unparalleled transformative restructuring of the high school science curricula. Teachers, educators, scientists, and officials of major organizations (National Academy of Sciences, AAAS, and the National Science Teachers' Association) have participated.

Leon Lederman currently holds the positions of inaugural resident scholar at the Illinois Mathematics and Science Academy, Pritzker Professor of Science at the Illinois Institute of Technology, and director emeritus of Fermi National Accelerator Laboratory. To the scientifically literate populace, he is known as the author of the highly acclaimed *The God Particle*. To others, he is the Mel Brooks of science.

On a personal note, Lederman is married to Ellen Carr Led-

Courtesy of Illinois Mathematics and Science Academy, Dean Kaus, Kaus Photography

Lederman signs *Portraits of Great American Scientists* with student author Maria Wilson at the book unveiling. Illinois Mathematics and Science Academy, November 14, 2001.

erman, a photographer. He has three children and four grandchildren; daughter Rena is an anthropologist, son Jesse is an investment banker, and daughter Rachel is a lawyer. When he is not improving the world with his zest for science, Lederman is an animal lover who enjoys horseback riding, downhill skiing, stargazing, baking bread, hiking, and telling jokes. At eighty years of age he is untiring and travels extensively to promote science education. Like a vintage wine, Lederman continues to age very well.

Leon Lederman's remarkable and unselfish generosity has distinguished his commitment to education. Although recipients of Nobel Prizes could decide to live and work within the boundaries of academia or research, his unparalleled passion for improving teaching and learning in science for all children has brought him into the schools and classrooms of America working directly with children and teachers, asking questions, probing understandings, and sharing his joy of science. His advocacy and tireless work to restructure and resequence the high school science curriculum has the potential if enacted to transform science education in our nation.

PARTNERSHIPS BETWEEN SCIENTISTS AND EDUCATORS

A Vital Ingredient in Science Education Reform

Marjorie G. Bardeen

INTRODUCTION

"Before It's Too Late" from the National Commission on Mathematics and Science Teaching in the Twenty-first Century (U.S. Department of Education 2000) is the latest in a series of reports on the state of K–12 mathematics and science education in the United States. The report calls for a "vigorous, national response that unifies the efforts of all stakeholders" to address the need for high-quality teaching. Among the stakeholders are higher education institutions and businesses. And among the suggested efforts to promote higher student achievement are working with area schools to identify needs for highly qualified teachers and making regular contributions of time, materials, and resources to enhance instruction in mathematics and science education in local schools. As early as 1980 Leon Lederman recognized this need and, as director of Fermi National Accelerator Laboratory (Fermilab), initiated efforts to make a difference. The twenty years of experience that Fermilab scientists have had working in partnership with K–12 educators can inform the scientific community as it takes up this latest call to action.

THE BEGINNINGS

Lederman was the director of Fermilab from 1979 to 1989. While he was at the laboratory, he and his colleagues completed a number of significant scientific achievements. For example, scientists on the E288 collaboration announced the discovery of a new particle named the upsilon and showed that the upsilon particle contains a b and an anti-b quark, thus revealing the third generation of quarks. In 1983 Fermilab scientists started the countdown for the world's first superconducting synchrotron, the Tevatron. At the time this accelerator, which remains the world's highest energy particle accelerator, was the world's largest project using superconducting technology and capped a ten-year effort to master the energy-saving technology for particle physics. Studying the production and decay mechanisms of the charm quark, the fourth quark of the Standard Model, was crucial in understanding the forces between quarks and how quarks combine to form composite particles.

But more important to this essay, Lederman used his prestige as a Nobel laureate and his position as a laboratory director for a bully pulpit to mobilize scientists to the cause of K–12 science education. There were no young students at Fermilab, so he began a precollege education initiative at the laboratory that grew to be an exemplar of a lasting partnership between scientists and educators to support science education reform.

Why would Lederman want research scientists to get involved in K–12 education? Beyond the genuine satisfaction many scientists get from teaching, there is an underlying need for preservation and diversification of the cadre of scientists. Scientists who rely increasingly on public support of their research need to raise the level of public scientific literacy and convince the public that investment in science is in society's best interests. Scientists need to inspire a new generation of scientists and, given the changing population demographics, they need to diversify the new generation. In order to stimulate the natural curiosity of students, efforts to encourage a lifelong interest in science must begin at the youngest levels. Whether these students become practicing scientists or scientifically savvy citizens, their support will be needed to maintain U.S. leadership in science.

Lederman began modestly by organizing a program called *Saturday Morning Physics*. He gathered together a group of ten senior Fermilab scientists and proposed a seminar program to be offered to as many as 100 high school students three times a year. Each three-hour seminar would include a lecture, discussion sections, and a lab tour. Each volunteer scientist would prepare one lecture; graduate students would conduct discussion sections and tours. Lederman sent letters of invitation to high schools within a radius of fifty miles of Fermilab and waited to see if any students would come.

Come they did, and their teachers came too! Students were stimulated by lectures on topics such as Accelerators and Detectors, Special Theory of Relativity, Leptons and Hadrons, and Cosmology. The young graduate students engaged the students in lively discussions and amazed them with a peek at the large, intricate accelerators and detectors at Fermilab. Based on the enthusiastic response of the students, Lederman felt that if the teachers could introduce modern physics topics into their courses, they could spark a similarly enthusiastic response. In talking with the teachers, Lederman realized that in general they knew very little about the physics and technology of Fermilab but were eager to learn.

Lederman dreamed of having a program for high school physics teachers as a companion to *Saturday Morning Physics*. Running a program for teachers would be more complex than the all-volunteer *Saturday Morning Physics* and would require plugging in to the world of staff development and school reform. Lederman recognized that the scientists ought to form a partnership with the educators. Scientists could contribute their knowledge of the subject matter and their understanding of how scientists work while teachers could contribute their understanding of kids and teaching and learning. However, to organize a program of significant size, a separate group was needed to serve as the bridge between the scientists and educators.

Lederman's timing was pretty good, as *A Nation at Risk* by the

National Commission on Excellence in Education (U.S. Department of Education 1983) came out in April 1983. Because of the critical nature of the needs in science education, he asked friends to form a nonprofit organization, Friends of Fermilab, to support K–12 science education programs at the lab. The money was raised for this first program for high school science teachers over just one weekend.

SCIENCE EDUCATION NEEDS

Much has occurred since 1983 to identify science education needs and to understand the magnitude of the effort that communities must make to meet those needs. Lederman's personal involvement and Fermilab's programs have grown accordingly. In a white paper published in 1998, Lederman and his colleagues speak eloquently about the need for strong science education for *all* students (Lederman et al. 1998).

Every graduating high school student of the twenty-first century must be ready and equipped to participate in and shape a society confronted with accelerating scientific advances, careers and jobs based on those advances, and increasingly wondrous technologies that impact our daily lives. Our nation's success in the twenty-first century requires that our citizens be scientifically literate and savvy. Our leaders, our parents, and our workers need to

- Be responsive to accelerating change;
- Work together to find measured yet creative solutions to problems that today are unimaginable;
- Anticipate the impacts of our actions;
- Communicate effectively;
- Maintain the balance and viability of our society and our ecology.

What should science education look like in order to help students achieve scientific literacy? Loucks-Horsley and colleagues (1990) and the National Center for Improving Science Education (1991) both present a vision and model for science education that has been termed *constructivist science*. The reports stress that students must experience in-depth engagement and that by doing so they develop content knowledge and scientific habits of mind. The National Center for Improving Science Education report (1991) is a synthesis of the significant reports issued since 1980 regarding the status of science education. These reports guide the direction of science education reform by asserting:

> How can educators turn kids on to science? The answer lies in helping children engage in experiences that require them to use scientific knowledge and processes as tools as they make sense of their experiences. This solution demands that the science classroom be transformed into an inquiry-based culture—a community of explorers—where curiosity, creativity, and questioning are valued, where resources and opportunities are made readily available, and where students can "work" like scientists engaged in the process of collective sense-making. Critical thought develops in this culture as new ideas are encouraged and where all of them—from the teacher, students, and textbook—are subjected to review and analysis by the scientific community of students. Children will come away from these experiences with the ability to use scientific knowledge to describe, explain, predict, and control their world.

Experts now realize that restructuring science education calls for a systemic approach. There are few exemplars to guide comprehensive change in school systems, and few of the current education leadership, lay or professional, have experience managing the change process. This task is bigger than the schools themselves, and we cannot expect them to work in isolation. Schools and districts must place a high priority on science education. They must find outside partners to help raise the level of community awareness and contribute to and support the needed change by sharing resources, expertise, and technology. Among these partners are colleges and universities, research institutions, and businesses and industries where scientists work.

ROLES FOR SCIENTISTS

As part of its basic commitment to science, Fermilab has pioneered education activities that engage young people as apprentice scientists and assist communities in providing exciting science activities that demonstrate sound scientific process and present content in line with current scientific knowledge. Fermilab offers programs in response to regional needs to sustain young people's interest in science and mathematics, to encourage young people to pursue careers in science and engineering, to revitalize the skills of current science teachers, and to maintain interest in the teaching of science as a career.

The involvement of as many as 200 members of Fermilab's technical staff, annually, in precollege education programs provides a vital interchange between scientist and teacher, between scientist and student. What do these researchers do? Their major roles include research mentor, content knowledge specialist, instructional development collaboration, and role model.

The role as research mentor is a familiar one for scientists; it's part of their work. Research groups are composed of scientists, postdoctoral fellows, graduate students, and, in some cases, undergraduate students. It does not take much imagination to bring secondary teachers, preservice teachers, or high school students into such a group. It may be that one such intern is added to the group, or in the case of larger groups with lots of support, one can find a "summer crew" of interns. For several summers one of the Fermilab technical groups included a physics teacher, two undergraduates (one from a historically black college), and a high school student from a public high school in Chicago. Each of these individuals had a firsthand opportunity to do science and understand such things as the importance of working collaboratively, learning from mistakes, repeating tasks when they don't work the first time, documenting work, and sharing results.

As content specialists, scientists give lectures and seminars, answer questions, and work in partnership with teachers to prepare materials for staff development programs and for classrooms.

This materials development provides some of the most interesting interchanges between scientist and educator. Whether figuring out how to share data online or creating a simulation of particle interactions with tennis balls and baby powder, the power of working together, sharing expertise, and respecting one another's contributions is truly exhilarating for teachers. Even in the simple opportunity to join scientists at the informal Director's Coffee held every afternoon, they realize, perhaps for the first time, that scientists respect them for their work as teachers, and they feel part of the scientific enterprise.

Just by meeting with students at the lab or in the classroom, scientists become role models. Most students have a stereotypic view of the "mad scientist" and are not familiar with a scientific workplace. Imagine the powerful impression a scientist can have when she talks with a group of young students about her work, her hobbies, her education; when she shares something from her work as a demonstration or hands-on activity. Imagine the impact she has on students traditionally underrepresented in science when she greets the students in her native language, Spanish. These seventh graders say it best:

> The most important thing I learned is that anyone can be a scientist. I saw people walking around in sweatshirts and jeans. Who knows? Maybe I can be a scientist.

> You cannot judge a book by the cover. Scientists come in all shapes and forms. Women, men, chemists, biologists, and physicists are all in the field of science.

> With most jobs you might say, "When is it ever going to be five-thirty?" But the scientists I talked to say, "Is it five-thirty already?"

LESSONS LEARNED:
SUCCESSFUL PROGRAM MANAGEMENT

How do scientists develop a partnership with K–12 educators beyond one of the obvious first steps, bringing people together? On an institutional scale like Fermilab, a small staff with previous experience and credibility in the program areas of interest should be assembled. This staff should be empowered to build the bridge between the scientists and the educators. Success or failure of the entire partnership may very well depend on the care with which these individuals are chosen. We can identify three other critical issues relevant for successful program management: institutional commitment, institutional approach, and participants.

The most crucial issue is institutional commitment. The partnership must have the director's or CEO's wholehearted, unstinting, enthusiastic support. Top administration must give visible support for the programs; reward staff participation, especially in job-performance assessment; sanction use of institutional facilities for education activities; and use its influence with other sectors to enlist and sustain their participation. Following closely behind institutional commitment is the need to create a partnership among equals. Rather than being driven by some scientist's "great idea," the proper attitude is one of equally excitable participants ready to argue and refine an idea on the basis of collegial partnership. Many may wish to participate, but some may not have the human relations skills to work with education programs, particularly if they are used to being experts. The idea of equal partnership must permeate the education program at all levels. Support should be available to those individuals who choose to become involved, and in some cases staff should be prepared to replace individuals who are not effective.

Building on the concept of partnership, Fermilab has developed a successful format for program development and implementation that includes two key components: conducting a needs assessment and establishing a program committee. Program development is guided by recommendations from an appropriate needs assessment, conducted with educators, community leaders, and Fermilab scientists. A typical needs-assessment workshop includes around twenty participants, lasts three to four hours, and covers a major program type or a new age group. It is important for scientists to listen to the education needs as perceived by the people who teach the students and employ the graduates. Scientists are not necessarily experts in precollege education. However, when they understand the local priorities for improving science and mathematics education, they can assess the laboratory resources and determine ways to work in partnership with educators to promote change.

We cannot overemphasize the importance of these needs assessments. All too often scientists from research institutions, colleges, or universities go to schools with all the answers. "We know if you do it this way, your problems will be solved." At our first needs assessment one high school science department chairman brought a list that covered seven legal-sized pages. Obviously, we could not meet all these needs, but we addressed some of them over the next eleven years through the Summer Institute for Science and Mathematics Teachers in which this department chairman played a major role.

After the needs assessment is completed and the results reviewed for program recommendations, a program committee, which is composed of local educators (master teachers, department chairmen, or instructional administrators) and a Fermilab scientist or two, works with Education Office staff to develop and conduct the program. The committee is responsible for such tasks as program announcement, curriculum development, participant selection, instructor selection, follow-up activities, evaluation, program reports, and dissemination. This is the foundation of the partnership that allows scientists to be scientists most of the time, teachers to teach most of the time, and all to join together with staff to support science education initiatives.

Strengths of this program-development model include

- Developing ownership of the programs among teachers by involving them from the beginning.

- Giving teachers leadership roles recognized both by their peers and research scientists.

- Establishing continuing communication channels between scientists and teachers.

- Utilizing expertise from various groups—teachers, science-education specialists, and scientists—with minimum interference with their regular job.

- Integrating national laboratory education programs with existing local programs.

The success of the Fermilab precollege programs led to the construction of the Leon M. Lederman Science Education Center, which houses an exciting set of interactive exhibits, a teacher resource center, a technology classroom, and a science laboratory. Students can discover the fascinating world of quarks and quasars through intriguing hands-on activities and multimedia kiosks that explore some of today's most amazing ideas and exciting scientific tools. In the Teacher Resource Center, staff members help teachers, students, and parents find instructional materials, gain awareness of the Internet, and connect with scientists. In the Technology Classroom, students and teachers use the Internet and receive Internet training, explore science with multimedia, and collect, analyze, and share data on computers. In the center's laboratory, students and teachers classify specimens from the Fermilab prairie and conduct a variety of experiments related to prairie fieldwork and to physics. Programs are offered in the following areas: Student Incentives and Opportunities, Teacher Preparation and Enhancement, Systemic Reform, and Public Awareness and Scientific Literacy.

CONCLUSIONS

Over twenty years after Leon Lederman sent his first letters inviting high school students to Fermilab, scientists continue to open the laboratory doors to K–12 teachers and students. Precollege education programs work at Fermilab because it is not business as usual. Teachers come to the world-class high-energy physics research laboratory for a unique opportunity to witness science conducted at the frontiers of human understanding and to learn from leading research scientists. Students have an experience in science that broadens and enriches their attitudes and develops their appreciation for science. Students see, perhaps for the first time, what the world of science is really like, and they like what they see! Fermilab cannot make change, but Fermilab can be an important catalyst supporting and nurturing change efforts of our education community.

These are the valuable lessons from the Fermilab experience. Scientists should

- Talk to teachers, go to schools, find out what science educators need.

- Get support from the highest institutional levels.

- Form a partnership with educators as equals; plan and offer programs collaboratively.

- Enjoy! If you are not having fun, regroup. This experience should be fun!

Whether in a small or large research group, whether at a research university, liberal arts or community college, business or research lab, we encourage interested scientists to work collaboratively with K–12 educators to improve science education.

REFERENCES

Lederman, L. M., M. Bardeen, W. Freeman, S. Marshall, B. Thompson, and M. J. Young. *ARISE American Renaissance in Science Education Three-Year High School Science Core Curriculum: A Framework.* Batavia, Ill., Fermilab-TM-2051, 1998.

Loucks-Horsley, S., R. Kapitan, M. D. Carlson, P. J. Kuerbis, R. C. Clark, G. M. Melle, T. P. Sachse, and E. Walton. *Elementary School Science for the '90s.* Andover, Mass.: The National Center for Improving Science Education, 1990.

The National Center for Improving Science Education. *The High Stakes of High School Science,* Andover, Mass.: The National Center for Improving Science Education, 1991.

U.S. Department of Education, the National Commission on Mathematics and Science Teaching for the Twenty-first Century. *Before It's Too Late: A Report to the Nation.* Washington D.C.: U.S. Government Printing Office, 2000.

U.S. Department of Education, National Commission on Excellence in Education. *A Nation at Risk.* Washington, D.C.: U.S. Government Printing Office, 1983.

THE ILLINOIS MATHEMATICS AND SCIENCE ACADEMY

A Commitment to Transformation

Stephanie Pace Marshall

I n a 1998 *Chicago Tribune* interview with Mary Catherine Bateson, anthropologist, author, and daughter of Margaret Mead and Gregory Bateson, Ms. Bateson was asked, "How does a parent today prepare a child for a future world that is difficult for that parent to imagine?"

Ms. Bateson replied, "Suppose you knew that your child would be part of a group that went to form the first colony on another planet, how would you prepare this child for life there? That's the kind of thing we should be asking ourselves about education. You can't prepare the child for the job market that will exist twenty years from now. So how can you build a curriculum that will shape an individual to be a pioneer in an unknown land—because that's what the future is" (Schreuder 1998).

Reflective educators have long been asking this question; but never has the need for a response, grounded in new insights about human learning and the transformation of the traditional structures of schooling, been more essential. The quality of our future is inextricably connected to our capacity for knowledge acquisition, knowledge generation, and continuous learning; these capacities will be the new measure of *wealth*, *wealth creation*, and *sustainability* in the knowledge era (Marshall 1997, 1998, 1999).

INTRODUCTION

This is a story.

It is the story of a relatively young institution, the Illinois Mathematics and Science Academy (IMSA). The story describes what the academy is doing *by design* to create a learning environment that develops and nurtures intellectual and creative talent, primarily in mathematics and science but also integrated with the arts and humanities. We want our students to be "pioneers in an unknown land" and we want to facilitate this process through the creation of an intellectually rigorous, imaginative, and reflective learning community that liberates the goodness and genius of all children, and invites and inspires the power and creativity of the human spirit for the world.

The Illinois Mathematics and Science Academy is an ongoing experiment in learning; a work in progress that seeks to understand and clarify the relationships between intentionally designed learning experiences that are competency-driven, inquiry-based, problem-centered, and integrative; the capacity of a learner to engage in deep disciplinary conceptual understanding, to create a more integrative and reflective mind, and to gain the ability and desire of learners to imagine and then work to create a compassionate and sustainable world that works for everyone.

It is my belief that there is a profound connection between the world we wish to create, the mind (and consciousness) needed to create it, and the learning experiences we create for children by design (Marshall 1997, 1998, 1999). It is this belief that drives us to engage in the work required to resolve two critical questions:

1. What is the nature of the learning paradigm that must ground our knowing for a new and sustainable global community?
2. What learning conditions enable this new paradigm of learning to become manifest in schools in order to make generative and integrated understanding more likely?

IMSA's story is our journey into the resolution of these questions.

IMSA'S CREATION AND HISTORY

Cofounded by Nobel laureate Dr. Leon Lederman and former Illinois governor James R. Thompson, the Illinois Mathematics and Science Academy was established by the Illinois General Assembly in 1986 as the United States' first three-year public residential learning laboratory for high school-age students highly talented in mathematics and science. Currently, 650 academically talented students from throughout Illinois attend the academy for three years (grades 10–12); admission is highly selective and there are no fees for tuition, room, or board. Since 1986, the academy has graduated more than 2,500 students representing 800 communities (towns/cities) from throughout Illinois.

The academy was also created to serve as a catalyst and laboratory for the advancement of teaching and learning in mathematics and science for all Illinois students and teachers. To serve this end, the Center for the Advancement and Renewal of Teaching and Learning in Mathematics, Science, and Technology (the Center@IMSA) was established in 1998 to design, develop, and deliver programs to other Illinois students and educators. This vibrant center serves approximately 2,000 students (grades 3–10) and 1,500 educators annually.

IMSA'S MISSION AND CHALLENGE

IMSA's mission is to ". . . transform mathematics and science teaching and learning by developing ethical leaders who know the joy of discovering and forging connections within and among mathematics, science, the arts, and the humanities. . . ." Working with internal students and with external students and educators, we have identified powerful and adaptive design principles for building the scientific understanding, cast of mind, and mathematical power of all students.

These design principles—*competency-driven learning experiences* that are *inquiry-based, problem-centered,* and *integrative*—

are fundamentally changing the ways that students and educators at IMSA engage in mathematics and science and with each other. Learning experiences designed with these principles enable students to acquire strong bases of disciplinary content knowledge (and the connections among these ideas) and skills; the ability to use the ideas, processes, and tools of mathematics and science to acquire and generate new knowledge; to apply knowledge to solve real world problems; and the predisposition to become apprentice investigators (Illinois Mathematics and Science Academy 2001).

Both our experience and our longitudinal research support the belief that all students will achieve at significantly higher levels of disciplinary and interdisciplinary understanding and will learn how to engage in the doing of real science, *if* learning experiences are intentionally designed to enable them to do so.

As a result, IMSA learning experiences enable learners to

- Direct their learning toward greater rigor, complexity, coherence, creativity, and integration.

- Increase their intellectual, social, and emotional engagement with and responsibility toward others around questions of genuine significance.

- Engage in collaborative, problem-centered, and inquiry-based learning that develops integrative ways of knowing, transdisciplinary connections, and curiosity.

- Participate in the world responsibly and fully because learners are fluent in multiple languages and ways of knowing—science, mathematics, poetry, music, nature, dance, arts, and humanities.

In the absence of these skills, children are learning-disabled in a knowledge- and information-driven world; they will not be able to invent the integrative mind required to advance the human condition in the twenty-first century.

CORE COMPETENCIES AND GOALS: BY DESIGN

IMSA's competency-driven curriculum is a coherent and integrated body of knowledge and skills that actively engages students with the understandings and insights of disciplinary and interdisciplinary knowledge. National standards (American Association for the Advancement of Science 1993; National Council of Teachers of Mathematics 1998; National Research Council 1996) in each discipline as well as the Illinois Learning Standards (Illinois State Board of Education 1997) were used as the initial framework for the development of IMSA's content standards (Illinois Mathematics and Science Academy 1999). The curriculum is also grounded in the institution's beliefs, philosophy, vision, and mission.

IMSA's philosophy is "to teach each child as if he/she is capable of significantly influencing life on the planet," our vision is "to liberate the goodness and genius of all children for the world," and our mission is "to transform mathematics and science teaching and learning by developing ethical leaders and integrative thinkers"; all three serve as reference for our curriculum and approach to learning.

IMSA's current work in curriculum and assessment design resides in the larger context of developing a coherent and integrated system of learning. This system will enable students to acquire, generate, and use knowledge for the world and is framed by four design principles:

- *Competency-driven* learning experiences enable students (1) to acquire strong bases of disciplinary content knowledge and skills, key ideas of the disciplines, and connections among these ideas; (2) to use the ideas, processes, and tools of the disciplines to acquire and generate new knowledge; and (3) to apply knowledge to address issues and to solve real-world problems.

- *Inquiry-based* learning experiences promote analytic thinking, knowledge generation and application, and meaning construction through mindful investigation driven by compelling questions that have engaged or have the potential for engaging the learner's curiosity (Illinois Mathematics and Science Academy 2000; National Research Council 2000).

- *Problem-centered* learning experiences are ones in which the learners grapple with complex, meaningful, and open-ended problems and work toward their resolution (Illinois Mathematics and Science Academy 2000; Torp 1998).

- *Integrative* learning experiences forge meaningful connections of concepts, constructs, and principles within and across academic subjects and real-world situations (Miller 1981).

These design principles focus instruction, assessment, curriculum development, and evaluation, as well as faculty professional development.

IMSA is striving to design a learning environment that provides a forum for risk, novelty, experimentation, and challenge and that gives power, time, and voice to student inquiry and creativity. This learning environment is grounded in IMSA's Standards of Significant Learning (SSLs).

STANDARDS OF SIGNIFICANT LEARNING— DEVELOPING THE INTEGRATIVE MIND

IMSA's curriculum is designed to enable students to demonstrate growth on IMSA's discipline-specific content standards as well as on our SSLs (Illinois Mathematics and Science Academy 1994). Together, these standards represent and provide evidence for the presence of those habits of mind that we believe contributes to integrative ways of knowing.

Specifically, the SSLs contain sixteen standards organized around five dimensions: developing tools of thought, thinking about thinking, extending and integrating thought, expressing and evaluating constructs, and thinking and acting with others. We expect these ways of knowing to broaden and deepen over time. (A complete listing of all IMSA standards can be found at www.imsa.edu.)

MATHEMATICS AND SCIENCE AS INTEGRAL TO THE HUMAN EXPERIENCE

As we continue to develop our program, we are acutely aware that decisions about curriculum, instruction, and assessment are fundamental decisions about the kind of minds we give our children the opportunity and invitation to create. If we are serious about solving the problems that plague us as a global community, we must invite our students to create the kind of mind that can identify and resolve these global human problems.

This is especially critical in mathematics, science, and technology education. Although mathematicians and scientists who have developed mathematical power and a scientific frame of mind understand the beauty, elegance, and symmetry of mathematics as a pattern language and know science to be a window to the wonders of the universe and the natural world, these understandings of and orientation to mathematics and science are foreign to most students. I believe this is because of the antiseptically rational and narrow (algorithmic) approach to mathematics and science that permeates most classrooms, emphasizing information accumulation while isolating students from the essential questions, perplexities, and wonders of the natural world.

It is no secret that it is possible to receive the highest score possible on a national standardized mathematics or physics exam and still not deeply understand basic concepts of the physical world. Although many high school students in the United States graduate with presumed disciplinary mastery, there is growing evidence to suggest that they also graduate with thinking that is characterized by stereotypes, misconceptions, unexamined assumptions, and rigidly held algorithms that inhibit their achievement of genuine and deep understanding (Schneps and Sadler 1987).

Mathematics is the universal language of science. It is a language of patterns and relationships as well as a discipline that explores relationships among abstractions. Students must perceive mathematics as part of the scientific endeavor. They must comprehend the nature of mathematical thinking, they must become familiar with how mathematical knowledge is constructed, and they must understand what drives mathematical inquiry.

Mathematics is a language of symmetry and interconnection, but students view it as linear and discrete. Mathematics is a form of abstraction, of symbolic transformation, and application, but students view it as a process of memorization and computation.

As children, we don't begin our exploration of relationships of the natural world in this way. At a deep and fundamental level, every child is born a scientist. Unfortunately, the mathematics and science taught in most of our schools diminish our natural capacity for inquiry and exploration. We must continuously rekindle students' inquisitiveness about the natural world by creating learning experiences that connect them to its wonders.

Within the framework of exploration and discovery in mathematics and science, students must be able to understand them as *languages* and as *ways of knowing*, whose knowledge base, symbol systems, concepts, and modes of inquiry and truth verification can enhance the understanding of other disciplines and other forms of knowing. Students must become multi-lingual; they must be able to translate and use the symbol systems of one discipline to understand the complexity of others. If they don't, they simply will not have the tools for knowledge generation in the twenty-first century.

Based on its beliefs, philosophy, vision, and mission, IMSA chose, by design, to create learning experiences in math and science that connect students to the elegance of mathematics and the wonder of scientific inquiry and discovery.

MATHEMATICAL INVESTIGATIONS

IMSA's core mathematics program is titled *Mathematical Investigations*. It is an integrated, four-semester course sequence, designed to place greater emphasis on multiple representations of ideas, reasoning, problem-solving, communicating, and connections among mathematical ideas and among mathematics and other disciplines. Students study concepts from all areas of precalculus mathematics including algebra, geometry, trigonometry, data analysis, and discrete mathematics, and they do so in an integrated, problem-centered, and collaborative manner. Students learn to use mathematics in a variety of intra- and interdisciplinary settings in addition to advanced studies in mathematics.

The goals of *Mathematical Investigations* are to

- Integrate topics from all areas of precalculus mathematics.

- Enable students to discover connections between and among algebra, geometry, and trigonometry concepts.

- Enable students to pursue calculus at the advanced placement level and to participate in other mathematics electives such as discrete mathematics, statistics, or courses in problem-solving.

- Enable students to question their assumptions about the learning and practice of mathematics, explore ways of verifying things for themselves, and apply mathematical knowledge to different contexts.

- Enable students to be responsible for accessing their own knowledge base rather than relying on review by the teacher.

This integrative, explorative approach to mathematical investigations allows the academy to incorporate multiple dimensions of mathematics, including data analysis and discrete mathematics (Illinois Mathematics and Science Academy 1999). *Mathematical Investigations* is about "putting different things together, not matching algorithm, ... the math becomes theirs ..." (Ron Vavrinek, IMSA mathematics teacher, personal communication, 1999).

SCIENTIFIC INQUIRIES

IMSA's core science program, currently being piloted, is titled *Scientific Inquiries* (Torp et al. 1999). It provides students with a rich and rigorous core science experience that explicitly infuses the power of student inquiry into the science curriculum.

The goals of *Scientific Inquiries* are to

- Immerse students in rich science content.

- Engage students in the identification and resolution of problems that integrate the learning and doing of science.

- Inspire students to continue their interest in and study of science and technology throughout their lives.

- Support students in becoming integrated learners characterized by complex thinking.

- Challenge students to demonstrate their genuine understanding of concepts through the use of multiple forms of assessment (IMSA is now in the process of developing the framework for an integrative learning portfolio that will represent the students' coherent integration and synthesis of all dimensions of their thinking over three years).

Specific inquiry modules have been designed to expose students to the unifying concepts and processes of science in a recursive way in order to enable them to progressively build awareness and knowledge and then to deepen that understanding for future learning in science.

The program is guided by the question posed from the student perspective: *How do I come to know the natural world and my place in it?*

The learning experiences in this two-semester sequence are initiated by phenomena or problems, providing a compelling reason to learn, investigate, and collect evidence in support of a growing understanding of scientific concepts and processes as they relate to significant issues and topics of interest or essential processes. As students delve deeper into their study, teachers take on facilitative, coaching, and mediating roles as well as modeling the investigative processes as colearners with students. The *Sci-*

entific Inquiries class is designed to be an authentic learning community, where the power of the idea to be understood is placed at the center of the inquiry. Both teacher and learner are engaged together in a communal and reciprocal process of discovery.

Essential to the philosophy of *Scientific Inquiries* is the engagement of students in learning science concepts and processes in a way that practicing scientists might encounter them, that is, through enticing phenomena or compelling problems. With the creation of inquiry-based learning environments, it is expected that students will discover the need for particular conceptual understandings or skills to be mastered before progress can be made on the broader problem or phenomenon under investigation. In practice, carefully designed instruction is offered to enable this understanding or skill to be learned or practiced within the context of the module or problem. Thus, the learning experience satisfies a desire to know when that desire is strongest.

Scientific Inquiries will support students in their desire and need to become responsible for their own learning by encouraging and promoting student inquiry within the context of assessment of competencies through dialogue, observation, intentional coaching, and professional guidance. Teaching strategies that foster student ownership and inquiry-based learning are used. As a result, students who are self-confident and self-directive science learners will develop and thrive.

STUDENT INQUIRY AND RESEARCH

The primary focus of the academy's program is to develop thoughtful inquirers, ethical leaders, and responsible stewards. While much of our instruction occurs through in-class seminars and laboratory experiences, significant learning and research takes place outside of the geographic boundaries of the Academy.

Through these learning experiences, students are actively engaged in knowledge generation as well as knowledge acquisition. They are collegial researchers and equal partners; they experience the joy of discovery and the disappointment of failure. They learn the power of following a question wherever it leads and of figuring things out. They are honored and respected for their talents and their questions.

Student Inquiry and Research (SIR) serves as an essential component of the academic program. Students work independently and collaboratively with their peers, practicing scientists, and scholars. These experiences foster their development as highly skilled and integrative problem-finders, problem-solvers, and apprentice investigators, allowing them to develop the skills required to succeed in the world of the twenty-first century.

The goals of Student Inquiry and Research are to

- Challenge students to engage in scholarly and scientific investigations, as well as artistic expression.

- Enable students to meet learning standards, generate knowledge, make connections, and develop a richer understanding of self, the world, and their place in the world.

- Enable students to investigate questions and plan disciplined creative work.

- Exhibit products of ethical research and inquiry.

- Build the capacity of students to design and execute self-directed learning experiences that develop the habits of mind of an integrative learner.

- Enable students to become skeptical inquirers who work at increasingly higher levels of independence, guided by professionals knowledgeable in their fields.

Participation in the SIR program encourages active student questioning, investigations, and presentation, situated in the context of concerns shared by a community of learners (Illinois Mathematics and Science Academy, Student Inquiry and Research Program 2000).

IMSA's SIR program provides a variety of research and inquiry learning experiences (Illinois Mathematics and Science Academy

2000; National Research Council 2000; Minstrell and van Zee 2000; Llewellyn 2002) for students to pursue compelling questions of interest, conduct original research in science and other fields, create and invent products and services, share their work through presentation and publication, and collaborate with other students and scholars throughout the world. Many students subsequently present their research in local, national, and international research forums; state and national competitions (Intel Science Talent Search Competition, Siemens-Westinghouse Competition, Junior Sciences and Humanities Symposium, Sakharov's Readings); and refereed scholarly journals.

Students participate in the SIR program through one or more of the following learning experiences:

Mentorship is an interactive off-site research partnership where students are paired with scientists and master scholars in museums, corporations, educational institutions, and research laboratories in the Chicagoland area. Research is conducted on-site one day a week in a variety of disciplines.

Inquiry is an in-depth study of topics developed in an individual plan of inquiry reflecting students' interests, guided by an experienced IMSA staff or faculty member.

IMSA courses offer students the opportunity for inquiry through the processes of investigation into a problem, problematic situation, model, or phenomenon.

ASSESSMENT OF LEARNING

Since the academy's opening in 1986, IMSA's faculty has developed and used multiple approaches for assessing student learning. Assessments have been linked to individual faculty's course objectives and, in some instances, to objectives for core courses taught by more than one faculty member. In addition to traditional approaches such as written examinations focused on acquisition of content knowledge, faculty members have developed and refined assessment strategies (Wiggins and McTighe 1998) that capture a wider scope of student learning indicative of IMSA's commitment to develop integrative learners—learners who are capable of acquiring, generating, and applying knowledge to real-world situations.

These strategies have included common assessments for some core curriculum areas, video assessment of foreign language proficiency, performance assessments, concept maps, reflective journals, projects, presentations, exhibits, and real-time feedback to instructors in the form of short writes.

As powerful as these assessment strategies have been in capturing a wider range of student learning relative to IMSA's commitment to develop integrative learners, they have not yet constituted a system for gathering and sharing information relating to development of deep conceptual understanding, how students represent their knowledge, and how students grow in disciplinary knowledge and skills. IMSA is now in the process of creating this system.

We envision that it will capture, monitor, and provide information about student learning for multiple audiences. For students, this information will help them make future learning decisions; for faculty, it will inform curriculum content and teaching practices; parents will be informed of their children's progress relative to learning expectations; for program and institutional leaders, it will inform program improvement, resource allocations, and stakeholders, for accountability purposes.

IMSA students historically perform at the upper 1 to 3 percent of the U.S. student population on traditional norm-referenced exams (Scholastic Achievement Test and the American College Test). In addition, their scores on the American High School Mathematics Exam (AHSME), an exam designed to identify and recognize talented mathematics students and potential mathematics Olympiad students) have annually ranked IMSA as one of the top five high schools in the nation. Furthermore, IMSA's own internal longitudinal study of AHSME scores indicates a small but statistically significant increase in scores after the introduction of IMSA's *Mathematical Investigations* program.

IMSA's work now is to design assessments that provide evidence of the growth of the "new mind" we are seeking to invite.

Such nontraditional assessments must be challenging, contextualized within authentic real-world problems, and generative.

SERVING ILLINOIS AND BEYOND

IMSA is positioned at the nexus of public education in Illinois as a laboratory for the advancement of mathematics and science learning and teaching. K–16 faculty, professional development specialists, and researchers use the IMSA laboratory and innovative learning and teaching strategies and materials to serve Illinois educators and the state education system through professional development, products for learning and teaching, and leadership in public policy. IMSA is committed to serving unmet needs of Illinois students and teachers in mathematics and science learning and teaching through a variety of programs (www.imsa.edu).

Programs for the students of Illinois include the following:

- *IMSA Kids Institute* serves students in grades 3–9 in hands-on enrichment programs that integrate concepts of science, mathematics, and technology with the humanities. The programs are designed and taught by IMSA students and other high school students who are trained by IMSA students and staff.

- *IMSA Excellence 2000* is an after-school enrichment program that (for up to two-and-a-half years) serves middle school students who are talented, interested, and motivated in science and mathematics, with special emphasis on students historically underrepresented and underserved in these areas.

- *Illinois Virtual High School (IVHS)* expands educational opportunities by delivering online courses to Illinois students. Through the IVHS, IMSA offers advanced courses in mathematics and science.

- *International Career Academy (ICA)* prepares high school students for academic and professional pursuits in international business by engaging a broad range of economic, political, and cultural issues impacting the diverse global economy. The student experience spans three summers and two school years and includes an internship with an Illinois multinational company.

Programs for the teachers of Illinois include

- *Problem-Based Learning Network* provides training and ongoing support to teachers as they use the research-based tools and methods of problem-based learning to improve student achievement in mathematics and science.

- *Bridges to Science Literacy* responds to the national imperative for high standards of knowledge and performance in mathematics, science, and technology. The American Association for the Advancement of Science's Project 2061 has created powerful tools for mathematics and science educational reform. This professional development program melds the success of Project 2061 in equipping educators to use state and national standards to improve student achievement with IMSA's experience in research, professional development, and creating standards-based learning environments. Additionally, this professional development helps teachers to update and deepen their knowledge in the areas of science, mathematics, and technology in order to keep these topics interesting and current for their students.

- *Twenty-first Century Information Literacy* trains and supports teachers, librarians and technology coordinators in using IMSA's powerful Internet Toolkit (http://toolkit.imsa.edu/locate/) as a resource to improve student learning.

Programs for the people of Illinois include

- *IMSA Great Minds Program* provides opportunities for Illinois educators, students, and the general public to learn from, interact with, and be inspired by great minds of our time, including Nobel laureates and other leaders in mathematics, science, the arts, and humanities through dialogues, seminars, and community lectures offered at IMSA, and at other sites in Illinois as well as online.

TWO STORIES

Why did IMSA choose the direction we did—to develop a competency-driven, inquiry-based, problem-centered, and integrative program that enables students to acquire, generate, and use knowledge for the world?

The answer was straightforward.

The structure of schools must be transformed because most schools' curriculum and instructional and assessment processes are antithetical to the principles of human learning. Schools are therefore not able to develop the integrative and collaborative mind (Marshall 1997, 1998, 1999).

THE CURRENT STORY OF SCHOOLING

I believe there is a powerful, often unconscious, paradigm of teaching and learning that is currently manifest in most schools. It is grounded in erroneous understandings about learning and the principles necessary to create learner-centered environments that genuinely liberate the learners' talents. I have characterized these assumptions as grounding the current story of schooling (see Figure 1), which is based on a culture of acquisition, independence, and competition.

The following assumptions underlie the current story of schooling:

- Learning is grounded in an epistemology that honors the objectively verifiable, the analytical, and the experimental; that views empirical observation as the most important skill; that believes that the acquisition of factual knowledge requires the disengagement of the learners' emotions in pursuit of objective truth; that believes that subjectivity endangers the pursuit of objective truth and that holds to the premise that there is no relationship between the knower and the known.

- Learning is an externally directed, passive, and linear process of acquiring information; false proxies (seat time, courses taken) are legitimate indicators of learning.

FIGURE 1—LEARNING "EVOLUTION"
CURRENT STORY
CULTURE OF ACQUISITION, INDEPENDENCE, AND COMPETITION

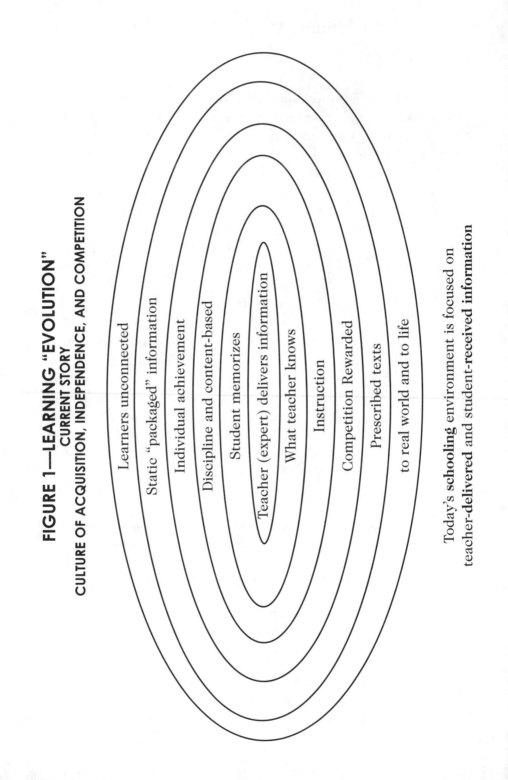

Learners unconnected

Static "packaged" information

Individual achievement

Discipline and content-based

Student memorizes

Teacher (expert) delivers information

What teacher knows

Instruction

Competition Rewarded

Prescribed texts

to real world and to life

Today's **schooling** environment is focused on
teacher-delivered and student-received information

- Intelligence is a defined and fixed capacity and is not learnable; analytical intelligence is the highest form of intelligence.

- Learning should be credentialed by the amount of time spent acquiring information; emphasizing authentic learning tasks that are complex, challenging, and novel and interferes with content and information acquisition.

- The purpose of schooling is to rapidly acquire information, cover content, and reproduce facts.

- Prior knowledge is unimportant and a detractor to future learning.

- Content segmentation, not concept integration, is the more efficient and effective way to learn a discipline.

- Rigorous and reliable evaluation of learning can only be objective and external; only that which can be quantitatively and easily measured is important knowledge; content coverage and information reproduction—not knowledge generation—are the most reliable indicators of learning.

- Competition and external rewards are the most powerful motivators for learning.

- Schooling represents a necessary rite of passage; what happens in school prepares one for life.

- Personal inquiry and the exploration of the learners' questions are not significant enough to take time away from the prescribed curriculum.

- Emotions, beliefs, and personal realities constructed from prior experiences do not influence and are not relevant to serious learning; they are only permitted if they do not significantly derail the curricular objective. (Marshall 1999)

All these assumptions point to the view of a passive and disengaged learner and a one-size-fits-all system that stifles our natural desire to learn. What we have come to understand at IMSA is that empowering and meaningful learning environments are created by intention and by design. We are fortunate that new insights from the neurosciences and cognitive sciences can now help us to create the conditions that invite human learning naturally in more effective ways than ever before. Sadly, however, despite these new understandings and new knowledge, most schools continue to be structured as if the mind functions best in a prescribed predictable, and sanitized environment. It is my belief, therefore, that as a result far more children are "school-disabled" than they are "learning-disabled."

THE NEW GENERATIVE STORY OF LEARNING

There is another story, however, and it is premised on trying to create conditions that foster the use and transfer of multiple symbol systems for learning; that connect mathematics and science with the arts and humanities; that require the consideration of ethical issues in the resolution of scientific problems; that foster interconnection and integration with real world issues; that immerse students in inquiry and messy ill-structured problems; that require collaboration; that inspire passion and curiosity; and that give students the opportunity to liberate their goodness and genius for the world.

I have characterized this commitment as the "new generative story of learning" (shown in Figure 2), which is based on a culture of inquiry, interdependence, and collaboration.

The following assumptions underlie the new generative story of learning:

- Learning is grounded in an epistemology that affirms integrative ways of knowing; that believes meaning and connections are constructed by the learner; that affirms the power of relationships and community in learning; that believes the learners' passion and love are essential for deep learning; that understands that relatedness and engagement are at the heart of learning and that there is a profound connection between the knower and the known.

FIGURE 2—LEARNING "EVOLUTION"
NEW STORY
CULTURE OF INQUIRY, INDEPENDENCE, AND COLLABORATION

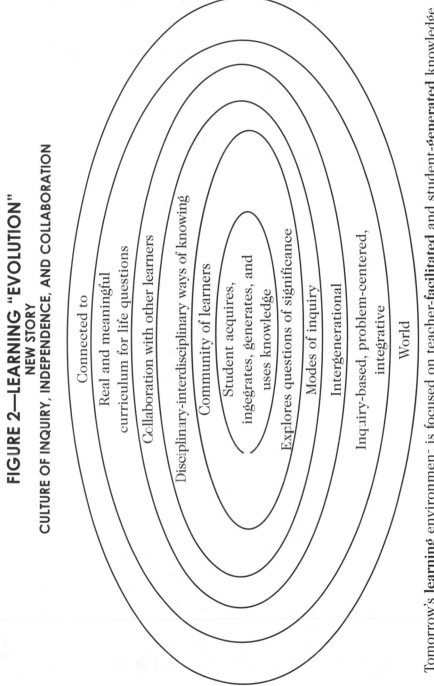

Connected to

Real and meaningful curriculum for life questions

Collaboration with other learners

Disciplinary-interdisciplinary ways of knowing

Community of learners

Student acquires, integrates, generates, and uses knowledge

Explores questions of significance

Modes of inquiry

Intergenerational

Inquiry-based, problem-centered, integrative

World

Tomorrow's learning environment is focused on teacher-facilitated and student-generated knowledge

- Learning is a self-directed, internally mediated dynamic process of constructing meaning through pattern formation.

- Intelligence is learnable, and potential and capacity for learning is inexhaustible and expanding.

- Learning is credentialed by demonstrations of understanding, at any time.

- The purpose of education is to develop understanding, wisdom, and the tools for lifelong learning through the reflective and often slow exploration of essential questions and through engagement in meaningful and challenging work.

- Prior learning is essential to future learning.

- Concept integration is the most meaningful way to understand the unity of knowledge.

- Rigorous and meaningful evaluation of learning must include qualitative evidence of understanding, be self-correcting, and be demonstrated in real-world settings.

- Collaboration, interdependence, and internal rewards are powerful motivators for learning.

- Learning is a continuous lifelong engagement; what happens in school is life.

- Personal inquiry and the exploration of deeply human questions are the means through which children acquire the knowledge and skills they need to construct meaning. The total engagement of the learner (intellectual and emotional) is essential for the construction of meaning.

- Engaged learning requires an intergenerational community learning together.

- Every epistemology gives rise to a pedagogy; how we teach is derived from how we believe people come to know. Enacting the new story of learning thus creates generative learning communities that invite students to become actively engaged in the development of their own mind. (Marshall 1999)

Recently one of our graduates returned to the academy from college, and I asked her, "How did IMSA prepare you for college?" Her answer was disturbing. "IMSA did not prepare me for college," she said. "IMSA prepared me for graduate school. In college you are taught to memorize, memorize, memorize. I am waiting to get to graduate school so that I can think again."

IMSA'S CONTINUING JOURNEY

Within the context of this new story of learning, there are a number of issues with which we are currently wrestling, and our hope is that our wrestling with them in the public arena, with an open invitation to our colleagues to engage with us, will not only open up the discourse but will inform the work we are all engaged in on behalf of the learning of our children.

Here are some of these issues and questions:

- Are there necessary tradeoffs between the amount of content a student acquires and the degree and level of his/her conceptual understanding?

- If a student excels on a standardized test, should the public assume conceptual understanding?

- How much content is essential for conceptual understanding and integrative thinking?

- Does interdisciplinary and integrative learning diminish or reduce a student's ability to acquire disciplinary understanding and disciplinary modes of inquiry?

Children are often bored in places called schools, not because we ask them to do too much but because we engage them in work that is far too small for their imaginations. IMSA is seeking committed partners to work with us in creating profoundly more meaningful ways to develop and assess deep understanding, especially in mathematics and science, so that all students have the habits of mind required to embrace with eagerness and confidence whatever lies ahead—as pioneers in an unknown land.

REFERENCES

American Association for the Advancement of Science. *Benchmarks for Science Literacy*. Project 2061. New York: Oxford University Press, 1993.

Illinois Mathematics and Science Academy. *IMSA's Learning Standards*. Aurora, Ill.: Illinois Mathematics and Science Academy, 1999.

———. *Inquiry and Problem Solving: Meaning-Making in Mathematics and Science*. Aurora, Ill.: Illinois Mathematics and Science Academy, 2000.

———. *IMSA Longitudinal Study of Graduates*. Aurora, Ill.: Illinois Mathematics and Science Academy, 2001.

———. *Standards of Significant Learning*. Aurora, Ill.: Illinois Mathematics and Science Academy, 1994.

———. *Student Inquiry and Research Program*. Aurora, Ill.: Illinois Mathematics and Science Academy, 2000.

Illinois State Board of Education. *Illinois Learning Standards*. Springfield, Ill.: Illinois State Board of Education, 1997.

Llewellyn, D. *Inquire Within: Implementing Inquiry-Based Science Standards*. Thousand Oaks, Calif.: Corwin Press, Inc., 2002.

Marshall, Stephanie Pace. "A New Story of Learning and Schooling." *School Administrator* 56 (1999): 31–33.

———. "Creating Pioneers for an Unknown Land: Education for the Future." *NASSP Bulletin* 82 (1998): 48–55.

———. "Creating Sustainable Learning Communities for the Twenty-First Century." In *The Organization of the Future*, ed. F. Hesselbein, M. Goldsmith, and R. Bechard, pp. 177–88. San Francisco: Jossey-Bass, 1997.

Miller, A. "Integrative Learning as a Goal in Environmental Education." *Journal of Environmental Education* 12 (1981): 3–8.

Minstrell, J., and E. H. van Zee, eds. *Inquiry into Inquiry Learning and Teaching in Science.* Washington, D.C.: American Association for the Advancement of Science, 2000.

National Council of Teachers of Mathematics. *Principles and Standards for School Mathematics.* Reston, Va.: National Council of Teachers of Mathematics, 1998.

National Research Council. *Inquiry and the National Science Education Standards.* Washington, D.C: National Academy Press, 2000.

———. *Science Education Standards.* Washington, D.C: National Academy Press, 1996.

Schneps, M. H., and P. M. Sadler. *A Private Universe.* South Burlington, Vt.: Annenberg/Corporation for Public Broadcasting Multimedia, 1987.

Schreuder, C. "Mary Catherine Bateson: Anthropologist and Author." *Chicago Tribune,* February 1, 1998, pp. 1, 4.

Torp, L., D. Dosch, D. Hinterlong, and S. Styer. *Scientific Inquiries: A New Beginning for Science at IMSA.* Aurora, Ill.: Illinois Mathematics and Science Academy, 1999.

Torp, L., and S. Sage. *Problems as Possibilities: Problem-Based Learning for K–12 Education.* Alexandria, Va.: Association for Supervision and Curriculum Development, 1998.

Wiggins, Grant, and Jay McTighe. *Understanding by Design.* Alexandria, Va.: Association for Supervision and Curriculum Development, 1998.

THE TEACHERS ACADEMY FOR MATHEMATICS AND SCIENCE

Lourdes Monteagudo

It was the fall of 1989, just after the School Reform Act of 1988 had gone into effect in Chicago. The white-haired Nobel laureate scientist, then Illinois governor Jim Thompson's science advisor, sat in front of the Deputy Mayor for Education of the city of Chicago and stated a problem. "All children are born scientists, but then they lose the enthusiasm to go into science," Dr. Lederman said sadly. "I have some good friends in the federal government that are very concerned about science education as a matter of national security, and I wonder how they could help Chicago improve the teaching of mathematics and science now that such a sweeping school reform effort has been legislated?"

I soon figured out that he was not an ordinary man. He was a genius. A man with a clue as to why children, especially poor minority children, lose their will to study mathematics and science so early. "Maybe it is something that happens to the children that is caused by some outside force that makes them lose their natural curiosity," he said with excitement as he scratched his white curls. "Most children in the United States are experiencing the same problem," he added. "Minority children are doing worse than others, but in international comparisons, even our best students struggle to keep up with the rest of the world. Maybe it is something that happens in the schools. There seems to be something fundamentally wrong with the way they are taught," he concluded. "Maybe with a few million dollars we could learn how to help."

This is when I knew that he was neither an educator nor a politician. Instead of barriers, he saw possibilities. He was not proposing a magic bullet but an experiment. He did not suggest that there was an inherent problem with the children or that the outside force was poverty, race, gender, or the neglect of parents. As a former elementary teacher and school principal in Chicago I knew that the "force" that Lederman sensed was poor teaching. Unfortunately, most elementary school teachers lack content knowledge and appropriate methodology to teach mathematics and science. This is a direct result of the lack of content rigor in the preparation of teachers and a pervasive misconception among policymakers that anyone can teach these subjects at the elementary level. As Lederman and I shared thoughts and ideas, the answer to his question began to crystallize. The idea of creating a center for the retraining of elementary school teachers was born.

A MOVE TO ACTION

Soon after that first dialogue, I found myself in Washington, D.C., visiting with Lederman's friends. Admiral Watkins, the Secretary of Energy of the United States, was particularly interested in the proposition that poor teaching was the problem, and that teachers could be the solution. If we could help those already in the classrooms gain the knowledge, skills, and attitudes they need to transfer the skills and excitement for the learning of mathematics and science to their students, we could accelerate the process required for a massive impact. In July 1990, Admiral Watkins and Lederman with the support of Chicago mayor Richard M. Daley held the ribbon-cutting ceremony for the Teachers Academy for Mathematics and Science. A widely representative board of directors, composed of leaders in education, science, business, and the communities was appointed to oversee the autonomous not-for-profit corporation.

As Lederman knew, and as more and more are now concluding, teachers cannot teach what they do not know. Most elementary teachers have been victims of the "force" themselves. Their own sense of wonder and curiosity was extinguished early on in their own school experience, never to be reignited, even after completing teacher preparation programs. Many elementary teachers never really understood algebra, never took a course in geometry, chemistry, or physics. The biology they remember was not the biochemistry of DNA, but something to do with animals and plants, which might be why so many elementary classrooms grow flowers in the spring and engage in lessons about dinosaurs. The focus on national and state standards and accountability gets lost in the daily routines as teachers cope with their anxiety over having to teach these subjects, by minimizing their importance in deference to reading and language activities, or by reverting to teaching the way they were taught.

No one knew for sure what it would take to teach underprepared teachers to be able to increase student achievement on standardized state tests. But it was clear that there was no time to waste and no room for disbelievers. The scale of the problem had to be addressed with a proportional solution. *The intervention had to be child-centered, school-based, and content driven. It had to offer the teachers the tools and help they need to make it all come alive in their classrooms. Retooling teams of teachers and not just enhancing their individual knowledge became the unique operational variable.* These simple but insightful tenets have become the basis of the academy's approach.

Lederman's deep-rooted belief in the capacity of all children to learn mathematics and science, his great tolerance for ambiguity, and his willingness to be an advocate for the importance of investing in the training of elementary school teachers have been the key to the model intervention that even after so many years is still ahead of its time.

THE TAMS APPROACH

After eleven years of work and analysis of achievement data, the academy has tested many of its hypotheses. We are convinced that most of the design elements are critical to school-wide improvement of student achievement in mathematics and science as measured by standardized tests. Now funded by the Illinois State Legislature, the academy has invested in the retraining of over 3,600 elementary teachers in 128 Illinois schools and has raised students' math and science scores among six of the lowest achieving school districts in the state including Chicago, Elgin, Aurora, Joliet, Cahokia, and East St. Louis.

The academy's program is designed to

- Engage teachers in a content review of elementary mathematics and elementary science guided by the vision of the national and state standards (120 hours over two years).

- Focus on the areas of the curriculum most neglected by the teachers due to their own lack of knowledge. In mathematics, number sense, geometry, measurement, and probability, and in science, the physical sciences, emphasizing the skills of data collection and analysis.

- Have teachers experience, as students do, the research-based methods that they should be using in their classrooms, such as cooperative learning, learning through inquiry, and the manipulation of concrete models and instructional materials.

- Offer teachers time to practice and transfer to students what they are learning with the help of in-classroom coaching offered by the academy's professional trainers.

- Make available free of charge classroom materials necessary to introduce and practice the manipulation of concrete models and materials to enhance understanding of concepts.

- Involve the principal and parents in the change process so that they can explore ways to support teachers as they change the culture of teaching and learning.

- Make linkages to the teaching of reading and language development.

- Create a critical mass of trained teachers in each school to increase the probability that students will be taught by a sequence of better-prepared teachers and that new teachers have better role models to emulate.

- Evaluate and report the results of the intervention using quantitative measures to determine the impact on teachers and students as measured by the state's standardized tests in mathematics and science.

Over the years, research in the improvement of mathematics and science education, and education in general, has slowly evolved to support the elements of the TAMS approach, as well as to accept the necessary costs, which once were described as outrageous. However, TAMS continues to be ahead of its time by having direct experience with the complexity of the change process in schools, the additional challenges that technology brings to the schools and teachers, the relationships necessary to make the intervention systemic and sustainable, and the accountability systems needed to keep the agency effective and accountable. TAMS experience suggests that

- Teaching teachers jointly, in groups that represent the grades within the cycle they teach (primary, intermediate, or upper), including special education and bilingual teachers, is an effective method to encourage teachers to learn and plan together in a heterogeneous learning environment.

- A critical mass of teachers from any given school must participate in the intervention if the outcome sought is a schoolwide average gain in the standardized state test scores.

- ° At least 70 percent of primary teachers must participate, with at least 75 percent attendance in sixty hours of instruction and fifteen hours of in-classroom coaching to show school-wide gains. When less than 70 percent of the teachers participate, the school-wide improvement in the students' achievement is overshadowed by the under-achievement of students in classrooms with nonpartici-pating teachers.

- ° The same is true for teachers in the intermediate grades. However, at least 120 hours of training in content and methodology is required to make a significant improve-ment in teachers' own learning and classroom practice due to increasingly complex content.

- Teaching teachers through the same methods that they are expected to use in their classrooms gives teachers a firsthand experience with the benefits of the methods and helps them identify and explore misconceptions in a safe, professional environment.

- In classrooms, support of teachers as they transfer the methods into classroom practice is essential for underpre-pared teachers to see the positive effects of new methods on their students. This motivates the teachers to try them again.

- Instructional technology must be introduced as a tool to sup-port the standards-based curriculum and not as a way of diluting the curriculum. Additional school-based training and support must be provided to develop the instructional skills of teachers.

- A strong textbook series, paired with a variety of classroom models and manipulatives, can help a school sustain the efforts over time.

- School districts with a centralized instructional focus and direction make greater gains than schools in districts where each school is "on its own."

- At least eight physical science modular units that emphasize variables that are basic to physical science and mathematics will significantly increase the math score of students in the middle grades.

- Providers of school-based staff development must have the capacity to deal with the scale of the intervention as well as the evaluation, instructional design, and quality assurance processes that will ensure consistency of service and tracking of results.

But after all is said and done, Lederman is never satisfied. "We have many more thousands of teachers to reach in our back yard, but why limit the reach?" he says. That is why he has been spending time in France, helping his friend Georges Charpak revolutionize the French school system with "hands-on" science and mathematics following the TAMS model. Following the First International Conference on primary Science Education in Beijing, China, other Asian countries are reaching out to him for assistance. Lederman knows no limits.

TRIBUTES

FROM LABORATORY GARDENS TO SOCIETY JUNGLES WITH LEON LEDERMAN

Georges Charpak

Our personal trajectories overlapped at one of the most beautiful towns of this planet, Venice, at the High Energy Physics Conference in 1959.

It was for me a decisive step. I had the luck, after the war, after a very poor training in physics but a respectable diploma as a mining engineer, to start in Joliot's laboratory at the Collège de France. His lectures on the history of nuclear physics were a beautiful introduction to science history. He was an artist in the operation of most detectors used at his time. Instinctively I tried to overcome my inferiority complexes in experimental techniques by investing my mind and my energy in the search for new detectors, with the blessings of the boss.

I came to Venice with an original new detector, useless but intriguing, which, for me, was to the future streamer chambers what some apes were supposed to be to the human race. Leon Lederman had illusions on my experimental capabilities to build high-voltage pulsed devices for the experiments he was planning at CERN, where he was to spend one year on sabbatical. He offered me a fellowship to join his group starting the measurement of the anomalous magnetic moment of the muon, and I spent the next thirty-five years at CERN! It was a challenging experiment, and I encountered remarkable personalities from whom I learned very much. After Lederman left he was replaced by Dick Garwin, from whom I discovered what it means to be a great experimentalist with an encyclopedic culture.

After three years the so-called g-2 experiment was completed and gave birth to several further generations of experiments aimed at drastically improving the measurement of the anomalous magnetic moment of the muon.

I decided to go back to research on detectors since the blooming field of high energy physics—with the birth of more and more energetic and more and more intense accelerators—was creating a strong demand for new detectors. While Lederman and Garwin were going to divergent activities, I kept a permanent contact with them, based on solid friendship and brotherly relations in many fields outside our profession.

I found in Lederman, who was jumping from one ambitious experiment to another, a permanent source of stimulating questions on new detectors adapted to the problems raised by the experiments he was planning. This inspired several important steps in the activity of my group, and ended up with an important collaboration among Fermi National Accelerator Laboratory (Fermilab), CERN, and SACLAY.

This collaboration led me to many visits to Fermilab, and on the occasion of one of the visits, Leon brought me to a school near Chicago where he was developing new teaching methods for elementary school children. I had the impression of having discovered America.

The practice of education that I saw there (hands-on science) was clearly one of the best ways to overcome the vast gaps existing in the world between educated and noneducated populations. I began lobbying to introduce this pedagogical method to France. I organized a meeting between Lederman and the Minister of Education, François Bayrou. We even went on television with him and in 1965 could organize visits to Chicago by big shots of the French teaching establishment and a few members of the French Academy of Science. They were convinced, and discovered at the same time the beautiful enterprise of the Illinois Mathematics and Science Academy in Aurora, which is a temple of excellence in the field of scientific education in high schools.

This started, for me, six years of intense activity in France. It encouraged me to look at the efforts undertaken by the National Science Foundation in a few centers in the United States in order to encourage, on a large scale, new approaches to elementary school education. I discovered Karen Worth, in Cambridge, Jerry Pine at Caltech, and encouraged my fellow citizens to be inspired by the American practice and to adapt it to our own conditions, which were in some aspects more favorable. The reward for these efforts is that in France the "hands-on" or "inquiry" approach is an official program of the Ministry of Education, which runs a body of 1.4 million employees. Six thousand teachers, at least, practice this method of education based on a scientific, experimental, active approach for the education of children. With many original developments due to the inventiveness of our teachers, many organizations became engaged in education, such as the schools for teachers at the university. With engineering schools participating in the development of the pedagogy, with the help of professors and many of their students, with local museums and also with the French Academy of Science, which carries great weight because of its prestige, we hope to be in a position to invite our friends from the United States to France to appreciate how we also contribute to this progress, which is, in fact, a challenge for all scientists on this planet.

For me personally, this has been the opportunity to exploit a prominent position originating from scientific activity to exert an influence in an important sector of our society. But this was not the only opportunity. My relation with Dick Garwin helped me to keep an eye on the fantastic developments in nuclear armaments on the planet. I participated with him in workshops on Star Wars, with people like Edward Teller, in Sicily at Erice, where another member of the "g-2"-experiment group, Antonino Zichichi, devoted part of his life to the development of a school tackling all planetary problems.

After receiving the Nobel Prize, I was invited to join a jury to elect Miss France! I refused for probably futile reasons. I was also invited to join a committee set up by the Prime Minister to discuss the future of the French nuclear arsenal. I accepted but readily realized how ignorant I was, as were many other participants. I discussed it with Garwin, and we decided to write a book providing solid data to the people who had the responsibility of making decisions on civilian or military applications of nuclear physics.

Our book was a best-seller in France. It was published in the United States in October 2001 and in paperback in December 2002.

What was striking for me was the level of superstition plaguing the field of nuclear radiation, where strident propaganda often replaces reasonable thinking. This led us to continue our cooperation and to our proposal to change the units used to evaluate the level of irradiation of population. I think that our presentation, submitted in June 2001 to the French Academy of Medicine, is worth presenting here. It is a step to clarify problems encountered by our society facing the effects of rapid scientific progress, and needing the participation of physicists in order to appreciate orders of magnitude in the relative hazards of various sources of energy.

For me, my recruitment at Venice by Leon Lederman has been the opportunity to enter a field of exciting physics, to reach a position where I can contribute to the scientific education of children, housewives, generals, admirals, and even politicians, and to cement solid friendships.

Decidedly, the "g-2"-experiment has been a success going beyond the respectable number of digits we added to the value of the magnetic moment of the muon.

LEON LEDERMAN'S PROSELYTIZING FOR SCIENCE

George A. "Jay" Keyworth II

In 1981 the U.S. high-energy physics community was at a watershed. While Fermi National Accelerator Laboratory (Fermilab) was thriving, there was no great vision of a next-generation facility to replace it, at least not one with much support. Some thought the next "turn" should go to the east coast, that is, to Brookhaven National Laboratory, in spite of its rather undistinguished facility proposal named Isobel. Since Stanford, with Stanford Linear Accelerator Center (SLAC), in the west and Fermilab in the Midwest had been the last major recipients, some thought it fair that Isobel be funded. Still others thought that the sociological challenges that CERN was confronting as a multinational cooperative venture were becoming manageable, and thus international cooperation was the only path to the future.

As the new science advisor, and one with particular interest in the issue, I arranged to meet one Saturday with the Department of Energy's (DOE) HEPAP (High Energy Physics Advisory Panel) on my own "turf," in the Old Executive Office Building. For me, the lasting memory of that meeting is a real tribute to Leon Lederman—nothing less but, unfortunately, also nothing more, since the Super Conducting Supercollider (SSC) never came to fruition.

Let me explain Lederman's role. First, it's important to put in perspective the psychology of that particular time. Recall that in the late 1970s and 1980s inflation had wreaked havoc on science budgets. Moreover, traditional American optimism was downtrodden, with our economy in double-digit inflation and the embarrassing and distressful predicament of American citizens held hostage in Iran, both issues seemingly without solution. Within the membership of HEPAP there was real doubt as to our national will to commit to any major, next-generation high-energy physics facility—which would surely cost billions of dollars. Isobel, however, might squeak through. Lederman was not really opposed to Isobel, as I recall, but he was focused intensely on what kind of facility would be required to respond to the questions that were then coming out of Fermilab and CERN. His Desertron, a global cooperation project, was designed to do just that. My response to HEPAP was to suggest that if HEPAP would endorse a largely U.S.-funded Desertron (which became the SSC) and abandon the lackluster Isobel project, in spite of it being Brookhaven's "turn," then I would do everything in my power to support the new project. And I did.

There are several things here that are worth remembering. One is that Lederman kept us focused on scientific excellence simply by pursuing his vision of what the next generation of researchers would require. He let nothing get in the way, and he was willing to do what was needed to make it happen. He was willing to stand up to his peers for something that was beyond compromise. And, to follow up on his vision, he was willing to communicate directly to the people who would fund the machine—the American public.

Prior to this time, virtually every major particle accelerator built in the United States had been justified using a strategy devised by E. O. Lawrence, which argued that particle accelerators were steps on the road to curing cancer. The SSC was too big and too costly (and cancer treatment had gained too much momentum on its own) to be able to rely on this otherwise tried-and-true strategy. As I discussed with the HEPAP members the need to sell such a project directly to the voters, Leon once again rose above the fray. It seemed to me that he simply viewed this as the simplest kind of challenge: If he was convinced that the SSC was a wise step for the country to take, then the least he could do was to try to explain to the people who would actually pay for it why that was so. In fact, Lederman actually enjoyed explaining, especially to laypeople, what most fascinated him about science and the pursuit of pure knowledge. Lederman then conceived a most remarkable thing. He wrote for *National Geographic* a most extraordinary article about the sheer beauty of understanding the nature of sub-atomic matter and why it was so important for us to continue on this path of scientific inquiry. And, as I still recall, it followed another fascinating article about exploring the ocean's depths.

Since World War II, basic research and the pursuit of pure knowledge has been supported in the United States in a manner that rivals the priority that art held in the Italian Renaissance. In that postwar era of respect for science, there were many heroes to capture the public eye, from Einstein to Fermi to von Neumann. Today there are too few heroes in science, and that support is weakening. The public rarely understands the difference between the quest for knowledge and technology and the application of that knowledge. And not enough scientists care to try to communicate the significance of that difference. But Leon Lederman is willing to do what it takes to earn "hero status" and, more important, to retain the public's support for pure science. In fact he may now be the best example I know of how to do it properly. And I am grateful to him for having reminded us.

LEON, FERMILAB, AND OTHER THINGS

Alvin Tollestrup

Leon Lederman became director of Fermi National Accelerator Laboratory (Fermilab) in 1979. The board of trustees really wanted him a year earlier, but they couldn't find him (he was basking on the beaches in Italy!). Then he said, "Okay, but I would like to try it for a year first." So he was director appointee for a while, but when he realized the incredible platform that came with becoming the salesman par excellence for physics, he gracefully accepted the job. He was also provided with a captive audience on which to practice his repertoire of jokes.

R. R. Wilson, the founder of the laboratory, was acutely aware of the impact that a great laboratory could have on the public. He realized that support of an esoteric subject like high-energy physics came from public funds, and that the public must be made to feel a part of the exciting quest. The laboratory was built to stand out. The central building, a sixteen-story structure on the flat lands of the central plains, can be seen from miles away. Its height and architecture beg for public exploration. The same kind of attraction is seen at closer range in the imaginative coloring and arrangement of buildings. The preservation of the original farmhouses, the buffalo herd, and the many sculptures add to the intrinsic interest of the place. Wilson laid the foundation.

This was the laboratory that Leon would direct and it was a perfect match of tool to man. The discussions of our future involved everyone. Leon's humor could defuse the most heated arguments, and his friendship with all people at all levels tapped into and mined the fantastic intellectual resources that reside at the laboratory. We flourished scientifically. But this discussion is for another day.

From the first, Leon involved the laboratory with the community, at both the local and national level. *Saturday Morning Physics* provided contact between high school students and working scientists. Courses for high school physics teachers followed. The summer program for students started. There was an attempt to start a multiuniversity institute that would attract scholars and provide an intellectual home for graduate students and postdocs near Fermilab. Lederman reached out to South America and Mexico and integrated them into his web. Government officials from congressmen and secretaries on down were invited to come and visit, and many, trapped by the warmth and humor of Leon, provided much-needed support in the future. His efforts spawned the Lederman Science Center at Fermilab, the Illinois Mathematics and Science Academy (IMSA) in 1985, and the Teachers Academy for Mathematics and Science (TAMS) in 1990.

Fermilab was used as a national (even international) stage to obtain support not only for physics but also for education in science. Fermilab provided much of the scientific personnel needed. But in every case, Leon obtained the best people from outside and supplied them with the resources they needed. His warmth, humor, and friendship and the many contacts mentioned above brought together teams that were extraordinarily productive. It was never "This is what you have to do," but "Let's find out what is needed and where can we get the resources." It was not a one-way street: The laboratory was also a much more exciting place.

I believe that IMSA embodies some of Leon's most lasting and significant work. Physics experiments can be clever and timely but in general they are soon superseded by new discoveries. However, setting in motion the organization of institutions that nourish the young minds of students and teach them the excitement of scientific discovery and the impact that science has on civilization has the potential to be far more important for our future than the discoveries we make in the laboratory. But Leon has excelled at both, and the world has benefited from his efforts. We can all wish him a happy 80th![1]

NOTE

1. It would be unfair to not mention Leon and his jokes. He really has fewer than is generally thought. Jokes come with three pieces: the prologue that introduces the characters, the situation, and finally the punch line. Long observation has indicated to me that he stores these pieces separately. And then it is like "mix and match" in a clothing store where many different outfits are available and can be selected and put together as the occasion demands.

LEON LEDERMAN

A True Friend of Science and Education

Alvin W. Trivelpiece

My experiences with Leon Lederman flow primarily out of my appointment in 1981 as the director of the Office of Energy Research for the Department of Energy (DOE). At that time Leon had only recently become the director of Fermi National Accelerator Laboratory (Fermilab). Even though I had served on a National Science Foundation committee with Leon, I didn't know him very well. As result of our respective jobs in 1981, that changed. I got to know Leon, his wife, Ellen, and Fermilab very well.

There were a few difficult moments, but on the whole, it was an exciting time during which I came to admire Lederman and to appreciate how difficult it is to run a national laboratory. Actually it is not difficult to be the director of a national laboratory; it is merely useless (JOKE!), as I was to learn later for myself. There are many forces that influence what is possible for a national laboratory, and there is the necessity to fight for budgets and programs. Nowhere in our academic or research training do we get any guidance on how to deal with the situations that seem to arise on an almost daily basis. Lederman cracked the code and was an excellent lab director.

Shortly after I started work at DOE, Lederman began the process of educating me about high-energy physics

(HEP). He wanted to make sure that I knew that it is more important than any other field of physics. He almost had me convinced. But then, subsequently, I would have to listen to the director or scientist or engineer from some other DOE lab as to why some other field of physics, chemistry, mathematics, computer science, etc., was clearly more important than HEP. So it went for the six years that I was director of the Office of Energy Research.

What caused Lederman to pull ahead of the pack was the close involvement that I had with him in the effort to gain funding for the Superconducting Super Collider (SSC). There were many friends and colleagues who provided help in making the case for the SSC. Even so, Lederman stands out as the one individual who always had some clever thing to suggest that might improve the chance that the SSC would get approved and funded. Most of his suggestions were excellent. Only a few were of questionable value.

At one point I had suggested that one of the problems was explaining to nonscientists what elementary particles are and why anyone should be interested. I asked Lederman if there were any videos or other instructional materials that explained the field in simple enough terms that I might be able to use to give to members of the cabinet, or even to President Reagan prior to any presentation requesting support and funding for the SSC. Sure enough! There was such a videotape. The subject of it was a judge who just happened to wander off the street into Fermilab (highly likely) to inquire about what went on in this interesting building that rose out of the prairie. This judge was right out of central casting and asked all the right questions. The tape had certain socially redeeming features, but something didn't seem quite right. A little interrogation of Lederman revealed that the "judge" was either a plumber or an actor. My concern over truth in advertising caused me to suggest that perhaps this wasn't the best vehicle to accomplish the intended purpose. Other similar helpful suggestions came along. In the end, the SSC was approved for construction. Many people helped achieve this goal; some more than others. Lederman was one of those who helped a lot.

Leon Lederman's abiding faith in the value of knowledge and his dedication to ensuring that such knowledge is passed along to the next generation is unique. He has gone above and beyond whatever might be expected in terms of using his powers of persuasion to get people to volunteer to teach courses, provide buildings, and so forth. He has used his selection as a Nobel Prize winner to further worthy goals of making it possible for more young people to have access to opportunities to learn about science and engineering.

As we have gone in somewhat separate directions the past few years, there haven't been regular opportunities to compare notes on how to advance the causes of improved opportunities for education and learning. I miss that.

Leon Lederman has more than earned this moment to bask in the light of recognition and appreciation for his many outstanding accomplishments.

EPILOGUE: OBSTACLES ON THE ROAD TO UNIVERSAL SCIENCE LITERACY

Leon M. Lederman

The wisdom contained in this book is awesome, the praise is fulsome, and my "response" task is gruesome. There is a common theme, the many failures of our education system, which has generated a rich rhetoric, such as "we have committed unilateral educational disarmament" (1983), or "by the year 2000, American students will be number one in math and science" (1989), or "before its too late" (2000) and "leave no child behind" (2002). The rhetoric that promises results is wrong for the simple reason that the rhetoric has never been followed by action.

After many frustrating but engaging decades trying to improve science education, I can now make a start in listing the obstacles to solving the problem of creating a superior twenty-first-century educational system.

Science, technology, and invention are often confused. Science is presented as a saving grace: a telephone, a cotton gin (converts more cotton to gin than one hundred men doing it by hand), a steam engine. The nature of scientific inquiry—the art of composing questions or of seeking answers to questions not yet asked—is missed. The distinctions and interdependence of science, technology, and invention are almost never a part of the curriculum—but they should be. Elements of engineering provide the student with a vital aspect of modern society. The transition from invention to science-based technology is another important feature of the continuous reshaping of our society.

The dazzling beauty of nature revealed by our developing understanding should be presented with enthusiasm. The sense of wonder at the simplicity of the laws of nature and the mystery of how mathematics seems to be embedded in these laws are important to convey. Here, the science teacher can find common ground with teachers of art, music, and literature. This suggests that periodic meetings of teachers across all fields of learning may reveal subtle and not-so-subtle connections. A literature of such connections exists and ought to be brought to the attention of high school teachers.

In the teaching of science to all students, not enough time is spent on process, on how science works, and on stories from the history of science, so that students will know how we know, how mistakes are corrected, the need for skepticism and openness, and the respect for curiosity as a drive for understanding.

Parents too often have little interest or time to pay attention to the quality of schooling their children are receiving.

Poverty, hunger, poor health, and the lack of good nutrition, health and dental care, and exercise—all make a child not "ready to learn."

Urban children living in the ghettos of our great cities too often do not have the family support, encouragement, and help that spurs the effort to learn. Too often peer pressure pushes in the wrong direction.

The road from home to school passes through neighborhoods that may have dangerous streets and gangs, only to arrive at decaying school buildings.

Diversity—that aspect of American society of which we are most proud—presents the classroom teacher with problems of students from a vast array of cultures and languages. Even if all the students were from a mountain village in the New Guinea highlands, the teacher would have to be aware that each student is an individual with personal strengths, with individual misconceptions, with different cognitive abilities, so that a one-size lesson does *not* fit all.

Schools are embedded in communities in which all members are "experts," most having spent a grand total of at least twelve years of their lives acquiring their expertise. Schools are then buffeted by political, social, cultural, and ethnic forces exerted by the community surrounding the school. One finds parent groups, soccer moms, community groups (UNO, Urban League, and so forth), teachers unions, and political cliques—all with very strong opinions. But all agree that the schools need to be protected from physicists.

Poor schools fail to provide the computers, the educational technology, and even some of the most basic educational materials.

The objective of the core science curriculum should be to excite the students, to imbue them with a scientific way of thinking that has validity and empowerment outside the fields of science.

The important topics of estimations, statistical inference, and probability are rarely offered in high school, but some sense of these disciplines would clarify issues and aid in decision making.

Meaningful, coherent, and seamless curricula in science do not exist. Colleges pay little attention to high school preparation and high schools do not stoop to see the middle schools, nor do middle schools pay attention to primary schools. The practice in 99 percent of U.S. high schools of introducing science via ninth grade biology is a nineteenth century obstacle to progress in science literacy.

A serious obstacle to allowing well-trained, well-prepared teachers to work their magic in their classrooms is the relatively "new bomb" called *high-stakes testing*. The need for accountability has mushroomed to unreasonable proportions, often confronting teacher, principal, and administrator with a conflict between professional advancement and the education of their students. Assessment is a crucial element in education, but it must be properly embedded in the educational process as a tool for the teacher and educational administrators. The present situation, unfortunately, exacerbated by the standards movement, must be replaced by kinder, gentler, and more constructive techniques.

In spite of an impressive specialized science education literature, there is no universal belief on the part of teachers, administrators, and parents that an education in science and mathematics is important to students who will not pursue careers in science. Teachers, guidance counselors, and parents continue to propagate the myth that women and some minority groups cannot "do" math and science.

For far too many students, the motivation for studying mathematics, science, European history, or even for staying in school after tenth grade is missing. Student mobility and teacher turnover are too often frequent and disruptive.

The career of "teacher" does not attract the best and the brightest students because the profession does not seem highly valued in our society, and because the overabundance of rules and the one-size-fits-all mentality stifles the creativity of the brightest of them. There is an endemic shortage of science and mathematics teachers. Teachers are poorly trained in subject content, but also in how students learn. The economic and social status of teachers must be drastically raised. Finally, the administrative requirements that distract from teaching serve to produce an alarming dropout rate. Taken together, there are problems in recruiting, training, and retaining good teachers.

Professional development in U.S. schools is nonexistent or inadequate. Teachers are not given serious time to prepare, to work with other teachers, or to seize mentoring opportunities.

Since we are on the topic of public science literacy, we should not ignore the colleges and universities. The old ethic in liberal arts colleges was to require a one-year science course, usually "rocks for jocks"—the famous Geology 101. Many universities do not require a serious science requirement; few achieve the level of the University of Chicago. There, two years of science with laboratory are required. One would think that this would be a minimum requirement for an educated college graduate of the twenty-first century. And while we are in universities, we should think about the professional schools. Think of the science-related issues that arise in the law: issues of the validity of fingerprints, DNA, and polygraphs; the issues of genetic testing and privacy. New laws are often generated by new technology, such as computer hacking. How can law schools *not* add a science requirement? The same arguments can be made for schools of journalism and business, and police academies. We live in a world in which technology modifies behavior in all directions. If the average citizen is not able to grasp the basic issues of just how the technology introduces new ideas, new constraints, new opportunities, how can he make informed decisions as a citizen, community member, and consumer?

We have so far listed obstacles in the organization, management, and curriculum of schools and colleges. However, "knowledge" about the world flows from TV, video, cinema, newspapers, magazines, and books (like this one)—collectively assembled under the rubric of "media." In my view a major obstacle to the achievement of science literacy is the general failure of media to convey the newsworthiness, the value, the excitement, and the significance of scientific happenings. Portrayal of junk science as news, play given to UFOs, and sensationalism of claims of the paranormal are properly lumped as antiscience. We recognize that the "F+" grade we give to media does include valuable exceptions, but these are drowned in a tide of neglect or worse—as exemplified by network TV.

The greatest of all obstacles, the height and depth of which frightens even physicists, is the resistance to change. To the question of whether this is worse among educators, education administrators, and all their sisters, cousins, and aunts, I can only shrug. Clearly, scientists welcome change: change of beliefs, change of tools, change of what you do and how you do it. But what about lawyers (who look back for precedents), politicians, police, and poets? The paradox is that schools must prepare twenty-first century graduates for change, driven by science-based technology.

My intention in starting this chapter was to list all the obstacles to creating better schools—schools appropriate to the twenty-first century—then outline a diabolically clever plan for overcoming all of these obstacles. However, creating the list was almost as hard as dialing long distance or blowing out the candles on my large birthday cake; I am exhausted.

Following the famous seventeenth-century mathematician Pierre de Fermat, I do believe there are solutions to all of the problems listed; in fact, guided by the contributions to this book by my esteemed colleagues, I have all of these solutions in hand, but there is not enough room in the margins to present them in this volume!

Contributors

BRUCE ALBERTS is president of the National Academy of Sciences and chair of the National Research Council, the principal operating arm of the National Academies of Science and Engineering. He is one of the principal authors of *The Molecular Biology of the Cell*, now in its fourth edition, considered the leading advanced textbook in this field and widely used in U.S. colleges and universities. Alberts is committed to the improvement of science education and helped to create City Science, a program for improving science teaching in San Francisco elementary schools.

MARJORIE G. BARDEEN is manager of the Fermi National Accelerator Laboratory Education Office. She has served on the board of trustees at the College of DuPage and was its chairman between 1990 and 1992. She served on the Board of Education of Glenbard Township High School District #87 between 1979 and 1985, and was its president from 1980 to 1985. Bardeen was the recipient of the 1984 Those Who Excel Award of the Illinois State Board of Education and was named the 1989 Outstanding Woman Leader in Education by the Suburban YWCA. She received the 1990 Max Bieberman Distinguished Alumni Award from University High School in Urbana, Illinois. Bardeen received a B.A. in mathematics in 1963 from the University of

Minnesota, and an educational certificate in mathematics in 1984 from Elmhurst College in Elmhurst, Illinois.

RODGER BYBEE became executive director of the Biological Sciences Curriculum Study (BSCS) in 1999 after serving four years as executive director of the National Research Council's Center for Science, Mathematics, and Engineering Education in Washington, D.C. As associate director of BSCS between 1985 and 1995, Bybee participated in the development of the National Science Education Standards. Bybee has been the principal investigator at BSCS for four NSF programs: an elementary school program titled Science for Life and Living: Integrating Science, Technology, and Health; a middle school program titled Middle School Science and Technology; a high school biology program titled BSCS Biology: A Human Approach; and a college program titled Biological Perspectives. Bybee has written widely. He is coauthor of a leading textbook titled *Teaching Secondary School Science: Strategies for Developing Scientific Literacy*. His most recent book is *Achieving Science Literacy: From Purposes to Practices*. In 1998 he was awarded the National Science Teachers Association's Distinguished Service to Science Education Award.

GEORGES CHARPAK is a member of the French Académie des Sciences and of the U.S. National Academy of Sciences. He has long worked at the European Center for Particle Physics in Geneva. He received the 1992 Nobel Prize in physics for his invention of electronic detectors of ionizing particles, used widely in physics, industry, and biology. He has long been committed to improving science education.

HOWARD GARDNER is the John H. and Elisabeth A. Hobbs Professor of Cognition and Education at the Harvard Graduate School of Education. He also holds positions as adjunct professor of psychology at Harvard University, adjunct professor of neurology at the Boston University School of Medicine, and chair of Harvard Project Zero's Steering Committee. Among numerous honors, Gardner received a MacArthur Prize Fellowship in 1981. In 1990 he was the first American to receive the University of Louisville's Grawemeyer Award in Education. He has been awarded twenty honorary degrees—including degrees from Princeton University, McGill University, the National University of Ireland, and Tel Aviv University on the occasion of the fiftieth anniversary of the state of Israel. The John S. Guggenheim Memorial Foundation awarded him a fellowship for 2000.

RICHARD L. GARWIN is Philip D. Reed Senior Fellow for Science and Technology at the Council on Foreign Relations, New York. He received his Ph.D. in physics from the University of Chicago and has worked in particle physics, with liquid and solid helium and superconductors, and extensively in technology. He is IBM Fellow Emeritus at the IBM Research Division, and also adjunct professor of physics at Columbia University. He has spent about half his time working for the U.S. government in technology and security, in fields ranging from the technology of nuclear weapons to arms control, satellite reconnaissance, and the global positioning system. In 1996 he received from the U.S. foreign intelligence community the R. V. Jones Award for Scientific Intelligence, and from the Department of Energy and the president the Enrico Fermi Award for his work with nuclear weapons and their control. He is a member of the National Academy of Sciences, the National Academy of Engineering, and the Institute of Medicine.

MARGARET JOAN GELLER is an astrophysicist who maps the universe. Her studies have shown that galaxies like our own Milky Way trace out vast patterns, the largest we know in nature. As we look deeper into the universe, we look back in time. Geller and her colleagues soon plan to map the universe as it was in middle age. She hopes to learn how the patterns in the nearby universe evolved. Geller has made two award-winning films about her work: *Where the Galaxies Are* and *So Many Galaxies . . . So Little Time*. Geller is a senior scientist at the Smithsonian Astrophysical Observatory. She is a member of the National Academy of Sciences and the American Academy of Arts and Sciences. She has received a number of honors, including a MacArthur Fellowship (1990–1995).

STEPHEN JAY GOULD (1941–2002) was the Alexander Agassiz Professor of Zoology and professor of geology at Harvard University, and curator of invertebrate paleontology in the Harvard Museum of Comparative Zoology. He was a recipient of the MacArthur Foundation Fellowship, served as president of the American Association for the Advancement of Science, and was a member of the National Academy of Sciences. As one of the most popular and well-known writers of our time, he made evolutionary biology accessible to the general public.

ELNORA HARCOMBE is associate director of the Rice University Center for Education and project director of the Model Science Laboratory Project. She holds a Ph.D. in neurophysiology from Yale University and has conducted research on small neural networks. Her current research centers on the ways in which people think and learn and on educational settings that foster thinking and learning in science. Harcombe has taught science and education at all levels—elementary, secondary, and university. She has been director of the Model Science Lab since its inception in 1989. This project has twice been the recipient of the Exemplary Partnership Award presented by the Texas Alliance for Science, Technology, and Mathematics Education.

DUDLEY HERSCHBACH has pursued scientific research and teaching for nearly fifty years, chiefly at Harvard University, where he received his Ph.D. in chemical physics in 1958. He is currently the Frank B. Baird Jr. Professor of Science at Harvard University. His research on the molecular dynamics of chemical reactions was awarded the Nobel Prize in 1986. He has taught many subjects in both undergraduate and graduate courses, including general chemistry for freshmen, for the past two decades his most challenging assignment. He is also engaged in several efforts to improve K–16 science education and public understanding of science.

MAE JEMISON is founder of BioSentient Corporation, a medical technology start-up; a college professor; and founder of The Earth We Share, an international science camp for students twelve to sixteen years old. Jemison, the first woman of color in the world to go into space, served six years as a NASA astronaut. She was the Area Peace Corps Medical Officer for Sierra Leone and Liberia between 1983 and 1985. Her undergraduate majors in chemical engineering and African and Afro-American Studies informed her ongoing commitment to the critical importance of universal science literacy. Jemison is a member of the Institute of Medicine, and inductee of the National Women's Hall of Fame and the National Medical Association Hall of Fame, winner of the Kilby Science Award, and in 1999 was selected as one of the top seven female leaders in a presidential ballot national straw poll. In her book *Find Where the Wind Goes*, she discusses growing up intending to be a scientist on the south side of Chicago. She appeared on an episode of *Star Trek: The Next Generation*, resides in Houston, and loves cats.

GEORGE A. "JAY" KEYWORTH II is chairman of the board of directors and Senior Fellow at the Progress & Freedom Foundation. From 1981 to 1985 he served as science advisor to President Reagan and, concurrently, as director of the White House Office of Science and Technology Policy. Prior to this, he was director of the physics division at Los Alamos National Laboratory. He is a Fellow of both the American Physical Society and the American Association for the Advancement of Science. He serves on the board of directors of the Hewlett Packard Company and General Atomics, as well as several emerging high-technology companies.

EDWARD W. "ROCKY" KOLB is head of the NASA/Fermilab Astrophysics Group at Fermi National Accelerator Laboratory. He is also a professor of astronomy and astrophysics at the University of Chicago. Kolb is coauthor of *The Early Universe* (with Michael Turner), the standard textbook on particle physics and cosmology. His book for the general public, *Blind Watchers of the Sky* (winner of the 1996 Emme award from the AAS), is the story of the people and ideas that shaped our view of the universe. Kolb is a Harlow Shapley Visiting Lecturer and Centennial Lecturer with the American Astronomical Society. He has been selected by the

American Physical Society and the International Conference on
High-Energy Physics to present public lectures in conjunction
with international physics meetings.

LAWRENCE M. KRAUSS is Ambrose Swasey Professor of Physics, professor of astronomy, and chair of the physics department at Case
Western Reserve University. He is an internationally known theoretical physicist with wide research interests, including the inter-
face between elementary particle physics and cosmology, where
his studies include the early universe, the nature of dark matter,
general relativity, and neutrino astrophysics. He received his
Ph.D. in physics from the Massachusetts Institute of Technology
in 1982, then joined the Harvard Society of Fellows. He is a Fellow
of the American Physical Society and of the American Association
for the Advancement of Science. Krauss is the author of more
than 180 scientific publications and six popular books, including
the national best-seller *The Physics of Star Trek*.

NEAL LANE is the Edward A. and Hermena Hancock Kelly University Professor at Rice University, and holds appointments as
Senior Fellow of the James A. Baker III Institute for Public Policy,
where he is engaged in matters of science and technology policy,
and in the department of physics and astronomy. Prior to
returning to Rice University in January 2001, Lane served in the
Clinton administration as assistant to the president for science
and technology and director of the White House Office of Science
and Technology Policy from August 1998 to January 2001, and as
director of the National Science Foundation. Lane has received
many awards and honorary degrees and is a Fellow of the American Academy of Arts and Sciences and a member of a number of
professional associations.

NORMAN G. LEDERMAN is currently professor and chair of mathematics and science education at the Illinois Institute of Technology and past-president of the National Association for
Research in Science Teaching. He is primarily known for his
research and scholarship on the development of students' and

teachers' conceptions of the nature of science. He has also studied preservice and inservice teachers' knowledge structures of subject matter and pedagogy, pedagogical content knowledge, and teachers' concerns and beliefs. Lederman has been author or editor of five books and ten book chapters, and has published more than one hundred articles. He is currently writing an elementary science teaching methods book.

SHIRLEY MALCOM is head of the Directorate for Education and Human Resources Programs of the American Association for the Advancement of Science (AAAS). The directorate includes AAAS programs in education, activities for underrepresented groups, and public understanding of science and technology. Malcom was head of the AAAS Office of Opportunities in Science from 1979 to 1989. Between 1977 and 1979, she served as program officer in the Science Education Directorate of the National Science Foundation. Prior to this, she held the rank of assistant professor of biology at the University of North Carolina, Wilmington, and has also worked as a high school teacher.

STEPHANIE PACE MARSHALL is the founding president of the Illinois Mathematics and Science Academy in Aurora, Illinois. Prior to this, she served as the superintendent of schools in Batavia, Illinois, and as a member of the graduate faculty at Loyola University. In 1992 Marshall became president of the Association for Supervision and Curriculum Development International, the largest educational leadership organization in the world. She was instrumental in establishing the National Consortium for Specialized Secondary Schools of Mathematics, Science, and Technology, and served as its founding president for two years. She also serves as a private consultant, keynote speaker, and writer on issues critical to educational transformation.

WALTER E. MASSEY is the ninth president of Morehouse College, the nation's largest institution of higher education for men. He has held a range of administrative and academic positions, including provost and senior vice president of academic affairs at the University of California, director of the National Science Foundation (appointed by former President George H. W. Bush), vice president for research at the University of Chicago, director of the Argonne National Laboratory, dean of the college and full professor of physics at Brown University, and assistant professor of physics at the University of Illinois. Active in a range of organizations, Massey has served as the chairman and president of the American Association for the Advancement of Science, vice president of the American Physical Society, past chair of the Secretary of Energy Advisory Board, and a member of the National Science Board. He was recently appointed by President George W. Bush to serve on the President's Council of Advisors on Science and Technology.

LOURDES MONTEAGUDO has served as executive director of the Teachers Academy for Mathematics and Science since 1993. Before joining the academy, and after a yearlong sabbatical researching private ventures in education at the John D. and Catherine T. MacArthur Foundation, she served as Chicago's first deputy mayor for education under Mayor Richard M. Daley. From 1984 to 1989, Monteagudo was the principal of the Albert Sabin School, which became a model for the restructuring of educational practices and was able to document evidence of student achievement gains to meet state standards, although serving a poor and historically underachieving community. Prior to her principalship, Monteagudo served as an elementary school teacher in the Chicago Public Schools.

JUDITH RAMALEY is assistant director of the Education and Human Resources Directorate, National Science Foundation (NSF). Prior to joining NSF, she was president of the University of Vermont (UVM). Before coming to UVM, she was president and professor of biology at Portland State University in Portland, Oregon. Ramaley has a special interest in higher-education reform and has played a significant role in designating regional alliances to promote educational cooperation, including the new Vermont Public Education Partnership that brings together K–12, the Vermont state colleges, and UVM into an alliance to promote K–12 partnerships.

She has contributed to a national exploration of the changing nature of work and the workforce and the role of higher education in the school-to-work agenda. She also plays a national role in the exploration of civic responsibility and the role of higher education in promoting good citizenship.

MELVIN SCHWARTZ is professor emeritus of Columbia University, from which he received both his undergraduate and graduate degrees. He spent seventeen years there before moving to Stanford University. He has served as chief executive officer of Digital Pathways, Inc., a company dedicated to secure management of data communications, and as associate director, high energy and nuclear physics, at Brookhaven National Laboratory. In 1988 he shared the Nobel Prize in physics with Leon Lederman and Jack Steinberger for the discovery of the muon neutrino.

SHEILA TOBIAS has made an art and a science as a curriculum outsider. Trained in history, literature, and politics, she has pursued the question, Why should otherwise able and ambitious undergraduates not choose mathematics and science? Her inquiry has resulted in six books on math/science education at college, and five others. Educated at Harvard/Radcliffe, she holds advanced degrees from Columbia University and eight honorary doctorates. She is a Fellow of the American Association for the Advancement of Science and the 2001 recipient of the Education Research Award of the Council of Science Society Presidents. In the 1980s she was a member of the American Physical Society (APS) Committee on Women and Physics and was a member of the APS Women in Physics delegation to the International Union of Pure and Applied Sciences in March 2002.

ALVIN TOLLESTRUP graduated from Caltech in 1950 and helped build a 1.2 GeV electron synchrotron, which was used to study photoproduction of mesons. He was an early National Science Foundation Fellow at CERN in 1958. He left his professorship at Caltech in 1977 and joined Fermilab, where he led the superconducting magnet development for the Tevatron. He was one of the leaders of the group that built the CDF detector that discovered the Top quark in 1995. He was awarded the Wilson Prize and received the National Medal of Technology in 1989. Caltech gave him the Distinguished Alumni Award in 1992, and he was elected to the National Academy of Sciences in 1996.

JAMES TREFIL is Clarence J. Robinson Professor of Physics at George Mason University. A recognized leader in the field of scientific literacy, he is coauthor of the *Dictionary of Scientific Literacy* and the widely used textbook *The Sciences: An Integrated Approach*. He is a Fellow of the American Physical Society and the World Economic Forum. He has written more than twenty-five books on science for the general audience as well as numerous magazine articles, and serves on advisory boards for a number of print and broadcast organizations.

ALVIN TRIVELPIECE served as the director of Oak Ridge National Laboratory from 1989 through 2000. Prior to this, he served as the executive officer of the American Association for the Advancement of Science (AAAS). He came to the AAAS from the U.S. Department of Energy, where he served as the director of the Office of Energy Research. Trivelpiece was corporate vice president at Science Applications, Inc.; vice president for engineering and research at Maxwell Laboratories; a professor of physics at the University of Maryland; and a professor at the University of California, Berkeley, in the department of electrical engineering. While on leave from the University of Maryland, he served with the U.S. Atomic Energy Commission as assistant director for research in the Division of Controlled Thermonuclear Research.

MICHAEL S. TURNER is the Bruce V. and Dana M. Rauner Distinguished Service Professor and chair of the department of astronomy and astrophysics at the University of Chicago. He also holds appointments in the department of physics and the Enrico Fermi institute at Chicago, and is a member of the scientific staff at the Fermi National Accelerator Laboratory (Fermilab). Turner received his B.S. in physics from the California Institute of Technology in 1971 and his Ph.D in physics from Stanford University in 1978. He is a Fellow of the American Physical Society and of the Academy of Arts and Sciences and is a member of the National Academy of Sciences.

Turner is a cosmologist whose research focuses on the earliest moments of creation. He is one of the pioneers of the interdisciplinary field that has brought together cosmologists and elementary particle physicists. His current research deals with the mystery of why the expansion of the universe is speeding up and not slowing down, and the dark energy that is causing the accelerated expansion. At the instigation of Leon Lederman and David Schramm, Turner and Edward W. Kolb established the Theoretical Astrophysics Group at Fermilab to pursue the deep connections between cosmology and elementary particle physics. They also wrote the monograph *The Early Universe*.

CHARLES M. VEST is the fifteenth president of the Massachusetts Institute of Technology. Under his leadership, MIT has developed new programs and organizational forms to meet emerging directions in research and education, including forging new kinds of partnerships with industries and institutions throughout the world. He is a recognized leader in bringing issues concerning higher education and research to broader public attention and to strengthening national policy on science, engineering, and education. A member of the National Academy of Engineering, he has served on the U.S. President's Council of Advisors on Science and Technology and is vice chair of the Council on Competitiveness and immediate past chair of the Association of American Universities.